BRITISH INFANTRY UNIFORMS Since 1660

To
J. A. F. B.
In Loving Gratitude.

BRITISH INFANTRY UNIFORMS Since 1660

by

MICHAEL BARTHORP

Colour Illustrations by Pierre Turner

BLANDFORD PRESS

Poole Dorset

First published in the U.K. 1982 by Blandford Press,
Link House, West Street, Poole, Dorset, BH15 1LL

Distributed in the United States by
Sterling Publishing Co., Inc.,
2 Park Avenue, New York, N.Y. 10016.

British Library Cataloguing in Publication Data

Barthorp, Michael
British infantry uniforms since 1660.
1. Great Britain. *Army*—Uniforms—History
I. Title
356'.186 UC485.G7

ISBN 0 7137 1127 2

Typeset by Polyglot Pte Ltd Singapore

Printed by Toppan Printing Co Ltd Singapore

Contents

Preface

The aim of this book is to trace the development of the dress of the British Infantry, Guards and Line, from the establishment of the Sovereign's standing army in 1660 to the present day, in peace and war, in temperate and in tropical climates. The term *dress* embraces not only uniform — full dress, undress and service or combat dress — but also the equipment worn by officers and men and, in more abbreviated form, their weapons. The treatment is almost entirely confined to the Regular Army, although the dress of the Reserve forces (Militia, Volunteers or Territorials) generally followed that of the Regulars, except for periods in the nineteenth century where the Volunteers were concerned.

The text is arranged by chapters in sections corresponding to the major dress changes. Each chapter deals with a century. The first section of each chapter is a brief consideration of the organisational and tactical developments in the century that follows. Except for the more recent periods, when modern conditions have required greater uniformity of appearance, both between officers, NCOs and men, and between regiments, each section deals first with the uniform and equipment of the bulk of the Line infantry, the rank and file, and then attempts to explain where this differed in the case of officers, sergeants and musicians. It also deals with those regiments which enjoyed special dress distinctions, such as Guards, Fusiliers, Highlanders, Light Infantry and Rifles. In order that the subject may be related to its correct context, the text begins with a Chronology of the Infantry's chief campaigns, followed by an Introduction which outlines the subject as a whole and summarises the period leading up to 1660.

When dealing with over three centuries' worth of uniforms worn by an arm of the service noted for its regimental idiosyncrasies, it is impossible, in one volume, to be as detailed as the subject deserves and it has therefore been necessary in the text to treat the subject in a fairly broad way. However, much regimental detail can be found in the Appendices. Even so, there are bound to be one or two omissions and inadvertent inaccuracies for which the author must crave the indulgence and forgiveness of any who feel themselves slighted.

Written descriptions of uniform are dry stuff, even to the most dedicated student, and indeed could be nearly meaningless without illustrations. The text is therefore complemented by a large number of contemporary paintings, watercolours, prints and photographs in black and white, and enhanced by coloured plates containing ninety-six figures, executed by the masterly hand of Pierre Turner from information supplied by the author. Thus, while the credit for the reader's pleasure in the attractive and soldierly aspect of these figures must be entirely due to the artist's skill, the blame for any errors must sadly rest upon the author. One further point about the plates should be mentioned. It is obvious to anyone with knowledge of soldiers and soldiering that, under active service conditions, uniforms and their wearers become dirty, often bloodied, and reflect the wear and tear of long campaigning. Much current military artwork attempts to show such effects, doubtless for sound reasons. Nevertheless, such touches must inevitably be largely speculative and their 'realism' can obscure details of the dress depicted. In the selection and execution of the coloured figures for this book, author and artist have decided to concentrate on the cut and colour of the uniform while attempting to retain the period flavour of the original pictures or photographs on which they are based. 'Mud and blood', it is felt, can safely be left to the reader's imagination.

For the first eighty years of this survey, 1660–1741, the evidence, both documentary and pictorial but especially the latter, is relatively slight. The scale of clothing for a soldier is documented, the details of what it looked like are much harder to ascertain. The problem is aggravated by the considerable licence enjoyed by regimental colonels over the clothing of their men, which therefore might alter, at least in detail, according to the whims of successive colonels. Furthermore, since the very existence of a standing army was, at best, only tolerated by the nation, and since military costume differed little in its cut from prevailing civilian fashions, there was scant incentive for artists to record the dress of military men, particularly that of the ordinary soldier. Officers' portraits from this early period are scarce and those that do exist can only be regarded as a reliable representation of the particular officer portrayed, since their dress, at this time, was much less regulated than it was later on. Not until George II diminished the colonels' powers by instituting a system of clothing regulations backed by royal authority, does the path of the uniform student become easier. The student is also aided by a growing increase in military portraiture and in illustration of ordinary soldiers' dress.

In the nineteenth century the regulations became more detailed, large numbers of military paintings and prints appeared, some reliable others merely fanciful, and by the time of the Crimean War, uniform was being recorded by the camera. In addition it was an age when military men noted down their experiences, particularly in foreign parts, in diaries, letters and eyewitness sketches, all of which occasionally yield valuable information about the costume actually worn in the field. Further evidence of this nature emerges from the work of war artists and correspondents sent out to India, Africa and elsewhere in response to growing public interest in the Empire, and the forces which safeguarded it, in the second half of the nineteenth century. At home, the demand for military illustration, much of it for boys' books but none the less valuable, grew apace in the 1880s and 1890s, resulting in the prolific output of such artists as Richard Simkin and Harry Payne. In the twentieth century, of course, there are few military activities which cannot be captured by the camera. For the latter half of this survey, therefore, the evidence is far more plentiful than for the first, but against that the dress changes have been more frequent and the variety of clothing worn more extensive. The seventeenth century soldier had the one suit of clothes he was issued with; his modern counterpart has a dozen different orders of dress.

Measurements have been given in Imperial — unless modern, where they have been given in metric.

In the course of researching this subject over a number of years, and in the preparation of this book, the assistance of many institutions and individuals has been sought. Among the former, the author gratefully acknowledges the help given by the staffs of: The Lord Chamberlain's Office; The Royal Library, Windsor; the National Army Museum; the Imperial War Museum; the Scottish United Services Museum; the Armouries, H. M. Tower of London; the British Museum, Department of Prints and Drawings; the Victoria and Albert Museum; the Scottish National Portrait Gallery; the Africana Museum, Johannesburg; the University of Otago, New Zealand; the India Office Library; the Ministry of Defence (Army); the Army Museums Ogilby Trust; Sotheby Parke Bernet & Co. Also the headquarters or museums of many regiments, old and new, in particular those of: the King's Own Royal Regiment; Royal Anglian Regiment; Somerset Light Infantry; West Yorkshire Regiment; Cheshire Regiment; Royal Welch Fusiliers; Royal Irish Rangers; Worcestershire and Sherwood Foresters Regiment; Royal Hampshire Regiment; Staffordshire Regiment; Dorset Regiment; Royal Berkshire Regiment; Durham Light Infantry; Queen's Own Highlanders; Royal Green Jackets. The author is also indebted to the following individuals who have helped in one way or another, either in a private or professional capacity: Douglas Anderson; Major J. A. F. Barthorp; A. F. H. Bowden; P. B. Boyden; Mrs Anne Brown; the late Marquess of Cambridge; Mrs Janet Cameron; W. Y. Carman; Colonel H. C. B. Cook; Major D. M. Craig; G. A. Embleton; Andrew Festing; D. S. V. Fosten; Miss Sheila Gill; Mrs. Marion Harding; R. G. Harris; Peter Hicks; Oscar and Peter Johnston; Major E. L. Kirby; Major C. W. T. Lumby, Royal Anglian Regiment; R. J. Marrion; John and Boris Mollo; Major Frederick Myatt; Lieutenant–Colonel R. W. T. Osborne; Major G. C. J. L. Pearson, 7th Gurkha Rifles; E. J. Priestley; D. W. Quarmby; J. H. Rumsby; W. S. Sampson; Miss J. M. Spencer-Smith; Charles Stadden; Mathew Taylor. A special debt must be acknowledged to all contributors, past and present, to that invaluable source of information, the Journal of the Society for Army Historical Research. Lastly the author must thank Pierre Turner for his friendly cooperation and skill with the colour plates, and Barry Gregory of Blandford Press for his encouragement and forbearance.

M. J. B.
Jersey, C.I. *1981*

Chronology of the British Infantry's Chief Campaigns

1661–1684 Tangier campaigns against the Moors
1672–1674 War with Holland
1685 Monmouth's Rebellion
1689–1691 Jacobite Rebellions in Scotland and Ireland
1689–1697 WAR OF THE GRAND ALLIANCE
1702–1713 WAR OF THE SPANISH SUCCESSION
1715/1719 Jacobite Rebellions in Scotland and England
1742–1748 WAR OF THE AUSTRIAN SUCCESSION
1745–1746 Jacobite Rebellion in Scotland
1755–1763 SEVEN YEARS WAR
1767–1769 First Mysore War (India)
1775–1783 WAR OF AMERICAN INDEPENDENCE (Including hostilities with France and Spain)
1780–1784 Second Mysore War
1790–1792 Third Mysore War
1793–1801 FRENCH REVOLUTIONARY WAR
1799 Fourth Mysore War
1803 First Mahratta War (India)
1803–1815 NAPOLEONIC WAR
1812–1814 American War
1806/1812/1819 Kaffir wars at the Cape
1814–1816 Gurkha War (Nepal)
1817–1819 Pindari and Mahratta War
1825–1826 Bhurtpore campaign (India)
1825–1826 First Burma War
1835 Sixth Kaffir War
1837–1839 Canadian Rebellion
1838–1842 First Afghan War
1839–1842 First China War
1843 Sind and Gwalior campaigns (India)
1845–1846 First Sikh War
1846–1847 First Maori War (New Zealand)
1846–1847 Seventh Kaffir War
1848–1849 Second Sikh War
1850–1853 Eighth Kaffir War
1851–1853 Second Burma War
1854–1856 CRIMEAN WAR
1856–1857 Persian War
1857–1859 Indian (or Sepoy) Mutiny
1858–1900 Nineteenth century campaigns on the North-West Frontier of India, including:
- 1858 Sittana Expedition
- 1863 Umbeyla Expedition
- 1868/1888/1891 Black Mountain Expeditions
- 1877 Jowaki Expedition
- 1895 Relief of Chitral
- 1897 The Pathan Revolt (or Tirah Campaign)

1859/1860 Second and Third China Wars
1860–1861 Second Maori War
1863–1866 Third Maori War
1863/1866 Fenian Raids (Canada)
1864–1865 Bhutan Expedition (N E India)
1868 Abyssinian War
1870 Red River Expedition (Canada)
1874 Ashanti War (West Africa)
1875–1876 Perak Expedition (Malaya)
1877–1878 Ninth Kaffir War
1878–1880 Second Afghan War
1879 Zulu War
1881 Transvaal (or First Boer) War
1882 Egyptian War
1884–1885 First Sudan War
1885–1889 Third Burma War
1896 Ashanti Expedition
1898 Second Sudan War
1899–1902 SOUTH AFRICAN (SECOND BOER) WAR
1903–1904 Tibet Expedition
1908–1939 Twentieth century campaigns on the North-West Frontier of India, including:
- 1908 Mohmand Expedition
- 1919/1921–1924 Waziristan Operations
- 1930–1931 Redshirt Troubles
- 1936–1937/1937–1939 Waziristan Operations

1914–1918 WORLD WAR 1 (or THE GREAT WAR)
1919–1921 Irish Rebellion
1919 Third Afghan War
1929–1937 Arab Revolt in Palestine
1939–1945 WORLD WAR 2
1945–1946 Counter-Insurgency in French Indochina and Dutch East Indies
1945–1948 Jewish Revolt in Palestine
1948–1960 Malayan Emergency
1950–1953 KOREAN WAR
1952–1956 Kenya Emergency
1954–1959 Cyprus Emergency
1956 Suez War

1957–1958 Aden Protectorate, Jordan and Oman Emergencies
1962–1964 Indonesian Confrontation
1964–1968 Aden Protectorate (Radfan) and Aden Colony Emergencies
1969 Northern Ireland Emergency

Introduction

The infantryman, a soldier who fights on foot, has always formed the bulk and backbone of an army. The bulk, because he is the cheapest type of soldier to maintain; the backbone, because, in the ultimate test of battle, though an enemy may be weakened, even irretrievably shaken by the action of horsed or armoured cavalry and artillery, it is the infantry which must eventually close with the enemy and, having overcome their resistance, hold the position so gained. Furthermore, it is only the infantryman who can proceed in terrain whose nature hinders, perhaps even prevents, the movement and operations of the cavalryman and gunner. It is only he who can competently perform in situations for which the mobility and firepower of his brothers-in-arms are neither appropriate nor desirable. Infantry can function on its own, albeit for limited periods on a battlefield, though for much longer periods in less warlike operations, but neither cavalry nor artillery can prevail without its assistance. Rightly the Infantry arm has been called 'the Queen of Battles'.

Of all armies, the British Infantry is in two ways exceptional. Firstly, it has enjoyed a unique regimental system which, despite changing methods of warfare, economic stringency and reduction of British power in the world, has survived essentially for some three hundred years and has given it much of its moral strength and prestige. Secondly, throughout its long history, it has experienced a greater variety of campaigning in more parts of the world than the infantry of any other country. Both these factors have had considerable bearing on the subject of this book, the dress of the British infantryman (which of necessity must include his weapons and equipment) over the last 320 years.

The most salient feature of the British infantryman's uniform has always been his scarlet or red coat. As will be seen, there have been exceptions to this, and in our more utilitarian age duller colours predominate, but even today it can still be observed in the full dress of the Foot Guards and, occasionally, on drummers and bandsmen at ceremonial parades. This fine, martial colour has been worn by other elements of the British Army, and indeed by some other armies, but its visual effect on enemies and allies alike has generally been to signify the presence of the British Infantry. Before Ramillies, Louis XIV exhorted Marshal Villeroi 'to have particular attention to that part of the line which will endure the first shock of the English troops'. When Villeroi observed the red ranks massing against his left, he reinforced accordingly from his centre — with subsequent catastrophe for himself. Nearly 180 years later, at Ginniss in the Sudan, the infantry were ordered to resume their red coats, the better to overawe the Dervishes; this was the last occasion when red was worn in action.

In the late Middle Ages, long before the notion of clothing an army uniformly became common, red was beginning to be recognised in Europe as the mark of an English soldier; either from the red St George's cross displayed as a national identification sign, or the use of coats of that colour (or russet) by sundry contingents sent to the French wars. The first standing or permanent troops of the Crown, the Yeomen of the Guard, formed by Henry VII, initially wore the Tudor livery of white and green, but in the next reign scarlet or red became customary.

During the British Civil War there were regiments on both sides in red, but it was not until the formation by Parliament in 1645 of Britain's first standing army, the New Model, that red clothing became uniform for all. The basic dress of the New Model infantryman consisted of a short red coat or doublet with white linen collar, tied with strings, falling over it at the neck, loose breeches gathered at the knee with ribbons, stockings (two pairs occasionally being worn at the same time) and shoes fastened with laces or ribbons. With this costume the musketeer wore a broad-brimmed high-crowned felt or leather hat; the pikeman an iron 'pot' or helmet, back-and-breast cuirass, iron tassets protecting the thighs, and stout leather gauntlet gloves. The aspect of a New Model regiment was undoubtedly much less uniform, particularly in the lower garments, than would be achieved later, but nevertheless a means of distinguishing different regiments was found by lining or 'facing' the red coats with contrasting colours which would show when and if the cuffs were turned back, or the inside of the skirts became visible. A costume similar to that of the New Model infantry can be seen to this day worn by the Pikemen and Musketeers of the Honourable Artillery Company.

With the Restoration of Charles II in 1660, at the beginning of this survey, Cromwell's army ceased to exist, save for one regiment, that of General George Monck, who presented it to the King he had restored to his throne. As 'The Lord General's Regiment', and within ten years 'The Coldstream Regiment of Foot Guards', this ancient corps provides the connecting link between the New Model and the Sovereign's Army from 1660 to the present. As such it is appropriate that the first of the colour plates should show a Coldstreamer of this early period. Charles II brought his own 'Royal Regiment of Guards' from Holland, also clad in red. This regiment, together with Monck's, formed the nucleus of the King's Infantry.

The 17th Century

BACKGROUND

In 1661 the King's Infantry stood at two regiments of Guards and two regiments of Foot. The first of the latter, the Royal Regiment (later 1st Foot), had been in the French service and could trace its origins to a Scots corps that had fought for Gustavus Adolphus in the Thirty Years War. The other (later 2nd Foot) was raised specially to garrison Tangier, which became a British possession following Charles II's marriage to Catherine of Braganza. Her dowry also included Bombay, for which another regiment was raised, but this was then transferred to the East India Company and did not return to the British Army for another 200 years, as the 103rd Foot.

By the end of the century the regiments had increased to three of Guards and twenty-nine of Foot, though Charles's Dutch wars and William III's struggles, first against the Jacobite risings on behalf of the deposed James II — who had himself increased the Army to counter Monmouth's Rebellion — and then against Louis XIV, had seen other regiments raised, but these had been disbanded when peace returned. This raising of regiments purely for the duration of a war would become common practice.

Command and virtual ownership of a regiment was vested in its colonel, whose name it bore while his command lasted. The basic infantry unit was the battalion, of which Guards and the Royal Regiment each had two; the others had one. A battalion's strength varied from some 500 in peace to 800 in war, divided into a headquarters and twelve companies. Each company was commanded by a captain with two subalterns, and consisted of pikemen and musketeers (in proportion of one to three at the Restoration, reduced to one to five by the end of the century); two or three sergeants and one or two drummers completed the company. Pikemen were chosen for their height and strength, the better to wield the 16 ft pike in defence of the musketeers reloading and at close quarters, particularly against cavalry. With the introduction of the bayonet, from 1683 onwards, which enabled the musketeer to defend himself, the pikeman's importance declined until pikes were finally abolished in 1706. The elite status the pikeman had enjoyed began to pass, from 1678 onwards, to a new type of infantryman, the grenadier, who formed in the battalion a thirteenth company without any pikes. Armed with musket, bayonet, sword, hatchet and three hand grenades, the grenadiers were picked men and expected to set an example to the rest of the battalion.

When a seventeenth century battalion formed line of battle, all the pikemen were usually concentrated in the centre, ready to form a *chevaux-de-frise* behind which the musketeers could retire if attacked by cavalry. The grenadiers were

Mid seventeenth century pikeman's armour of 'pot' helmet, cuirass and tassets.

usually divided between either wing, with the musketeers on either side of the pikemen in five or six ranks, though the frontage of the battalion could be doubled by forming three ranks. Fire was delivered by ranks in turn, each countermarching to the rear to reload.

From 1685 an English and a Scots regiment of Fusiliers (later 7th and 21st Foot) were added to the Infantry. First formed to protect the artillery train on the line of march, this role was subsequently abandoned and the two regiments became ordinary infantry, though retaining their designations as an honorary title, which in later years would be granted to other regiments as a reward for distinguished service.

DRESS 1660–1700

A painting probably showing the 1st or Royal Regiment of Guards depicts them accompanying Charles II on his departure from Holland. The musketeers have iron helmets, hip-length buff coats with the sleeves hanging loose revealing a red doublet underneath, capacious red breeches, and stockings. This costume which held echoes of the Civil War period did not last long, for civilian fashions changed after the Restoration and military dress generally followed the same style.

A Royal warrant of 1678 laid down the dress of a soldier as: 'A Cloth Coat lined with bayes [baize]; one pair of kersey breeches; two shirts; two cravatts; one pair of shoes; one pair of yarne hose; one hatt edged and hatt band; one sash; one sword and belt.' Although there were minor differences between the dress of the Guards and the Foot, as there were between individual regiments of Foot, the appearance of both was broadly similar. Indeed, it would remain so for many years; for not until the 1830s would the uniform of the Guards differ markedly from the uniform of the Line.

The hat was black, broad-brimmed and lower in the crown than previously. For officers it was of

Musketeers, officers and a drummer, possibly of the Royal Regiment of Guards, 1660. Detail from a painting showing Charles II leaving Holland.

beaver or velvet, felt for the men, with ribbon and rosette or bow round crown, officers having plumes until 1685. The brims were edged with lace or galoon, which for the musketeers of the 1st Guards, was silver in 1685, gold for the Coldstream Guards; Foot regiments usually had white. Hats were turned up in front or at one side, later on two sides and eventually on all three. Pikemen continued to wear the iron pots of the Civil War but later, probably around 1680–1685, they adopted hats that were similar to those of the musketeers.

Since the hat was inconvenient for grenadiers when they were slinging their muskets prior to handling their grenades, they were given bag-caps which were usually red, bound round at first with fur but later with cloth in the coat facing colour, with a stiffened upright front of varying shape. This made them more imposing. The fronts of these stiffened caps also provided a place to display the Royal cypher, in the case of Guards regiments, or the colonel's crest. In 1684, for example, the Royal Regiment had caps with white fronts to match their facings and a crowned lion's head thereon. At the Coronation of James II, the herald, Francis Sandford, described the grenadier caps of the 1st Guards as 'red cloth lined with blew shaloon, and laced with silver galoon about the edges: and on the frontlets (which were very high and large) was embroidered the King's Cypher and Crown.' The Coldstream had their caps 'lined and faced with blew shaloon and laced with gold galoon, and imbroidered on the frontlets with the King's Cypher.' A portrait of a 1st Guards grenadier officer, c 1685, shows the frontlet to be almost square, with rounded corners. Fusilier regiments had similar caps to the grenadiers.

In a drawing by Hollar of troops at Tangier, the officers are wearing small grey hats with much narrower brims than was customary. Another exception to the normal hat was seen in the Earl of Argyll's Regiment, the first to be raised in the Highlands, though disbanded in 1698. Otherwise clothed as English infantry, its men wore the blue Scots bonnet on its formation in 1689.

Soldiers' hair was worn long and untied in the prevailing civilian fashion. In the case of officers, the long curled locks spreading across the shoulders, which can be seen in portraits of the time, were usually, but not invariably, wigs. The early years of Charles II's reign saw a vogue for small moustaches but by James II, all ranks, with the occasional exception of some grenadiers, were clean-shaven, and so they would remain for the next hundred and seventy years.

From 1670 onwards, the falling linen collar at the neck lost favour, being replaced by a neckcloth or cravat with loose ends which, for officers, was often of lace embroidered with gold or silver, for men, of an inferior material known as sleazy.

The red worn by the New Model Army continued as the coat colour for the King's Army. However, the hip-length doublets lengthened; becoming sleeved waistcoats when an almost knee-length coat, variously called a gown, cassock or surtout, began to be worn over them. The latter became the soldier's outer garment. These had no

Plate 1:1660–1700

1. Pikeman, Coldstream Regiment of Foot Guards, c 1670. 2. Musketeer, 1st Regiment of Foot Guards, 1686. 3. Company Officer, Bellasis's Regiment, 1692 (later 22nd Foot). All three figures wear the long coat, waistcoat, knee breeches and stockings which, subject to numerous modifications, were used to clothe the infantryman for some 140 years. The hat, which turned up on one or more sides, was used for a similar period. Early in Charles II's reign, the pikemen's coats of the Foot Guards differed from the musketeers' coats: the 1st Guards' were silver-colour, faced light blue, with white sash fringed blue; the Coldstream's were reversed from the regiment's red, faced green, with a white sash fringed green (Fig. 1). By the reign of William III the pikemen's armour had been discarded and their dress assimilated to the musketeers' dress. Fig. 2, based on Sandford's description of James II's coronation, has a collar of cartridges (bandolier) and is armed with a sword in addition to his black-painted flintlock, a weapon which was just beginning to replace the matchlock. Coldstream musketeers at this date were similarly dressed, but with gold-laced red-ribboned hats, red breeches and stockings. Officers' dress frequently lacked uniformity, though the crimson and grey clothing of Fig. 3 was ordered for all officers of that regiment. Commissioned rank is indicated by the gorget, sash and half-pike.

1. Pikeman, Coldstream Guards, 1670.

2. Musketeer, 1st Guards, 1686.

3. Officer, Bellasis's, 1692.

collars and initially hung loose and open. When the sword belt moved from the shoulder to the waist, they acquired buttons down to the hem and became more waisted, though most of the buttons above and below the waist were left unfastened. The skirts were slit at the sides and back and had pockets on the fronts. As in the New Model, the coats were lined in contrasting colours. The choice of colour was at the discretion of the colonel and so might alter when the regiment changed hands, or a quantity of cheaper cloth became available. For example, in 1689 Hodges's Regiment wore red coats lined red but two years later they were lined white. However, as the Royal livery was red and blue, facings in the latter colour became customary for Royal regiments, and surviving inspection returns generally show a certain continuity of facing colour for the remainder. Since the cuffs were now quite deep and permanently buttoned back, the facing colours were clearly displayed. At first only the 1st Guards' coats were in the Royal livery, the Coldstream's being faced green, changing to blue in 1685, and the Scots Guards white, becoming blue later. For some years the pikemen's coats were in reversed colours to the musketeers' (see Fig. 1, Plate 1), a practice followed by drummers of non-Royal regiments. This was abandoned for pikemen by the end of Charles II's reign, and their coats became the same as the coats of the musketeers. The body armour used in the Civil War continued for a while but the tassets were dispensed with soon after the Restoration, and the cuirasses at the same time as the pots. Stout buff leather gloves remained an essential part of a pikeman's clothing, to protect his hands from splinters on the ash poles of the pikes.

The coats of the grenadiers were made more distinctive by reinforcing the buttonholes and sometimes the seams with worsted tape or 'lace' loops, occasionally with tufts at the end. To make it more decorative, the lace acquired coloured lines or patterns of regimental choice. In 1689, for example, the grenadiers of Lord Bath's Regiment (later 10th Foot) had red and white loops, while Lord Cutts' Regiment (disbanded 1699) had black and white. It would appear that some sergeants and corporals, certainly in the 1st Guards in 1682, also had laced coats, though the type of lace is uncertain.

Bath's Regiment, when raised in 1685, was clad in blue coats, a colour adopted for several regiments raised by William III, while Tiffin's Regiment (later 27th Foot), raised at Inniskilling in 1689, initially had grey, as did one or two other regiments. Before long, however, the ubiquitous red coat became the rule.

Infantry officers, c 1670. From right: a captain, a lieutenant and an ensign; pikemen behind. From *Military Discipline or the Art of War* (1689).

The sashes mentioned in the 1678 warrant were in varied colours according to regiment. Their use would appear to have been mainly confined to sergeants and pikemen, although musketeers may have worn them before the sword belt was transferred to the waist.

Despite red coats and waistcoats being common to all (other than the exceptions mentioned), breeches and stockings varied between regiments, sometimes matching each other, sometimes contrasting. The facing colour was often repeated in either or both. For example, Lord Hastings' Regiment (later 13th Foot) being faced yellow had matching breeches but grey stockings; Cornewall's (later 9th) faced orange, had grey breeches and white stockings. Blue, grey, yellow and white

William III's Infantry, 1694. Grenadiers can be seen at the left rear, pikemen on the right. In the foreground officers salute with half-pikes and Colours, behind them are drummers and musketeers. From *The Funeral of Queen Mary*.

seem to have been the most common colours for breeches and stockings, but red and buff were used by some regiments, while Kirke's Regiment (later 2nd Foot) wore green in 1686. Breeches were cut less full than prior to 1660 and were tied below the knee with cloth ribbons, covering the stocking tops which came up over the knee. As in the Civil War, a second thicker pair of stockings were worn on service to protect the finer pair below; the former were often pulled up over the breeches. The low shoes worn at the Restoration acquired an upstanding tongue and by 1688 their ribbon or rosette fastenings were replaced by a buckle and strap.

As regards weapons and accoutrements, infantrymen of all ranks and all types carried a straight sword with brass or iron-hilted guard and leather scabbard, slung from a frog attached to a buff leather belt. As mentioned, this was worn over the right shoulder early in this period, transferring to the waist around 1685. From about 1687 the straight sword was probably superseded by the hanger, with shorter, slightly curved blade and knuckle-bow. However, this is uncertain as the hanger was frequently called a sword in contemporary documents. When the bayonet was introduced, a separate frog was stitched on to the belt to the left of the clasp. Grenadiers had an additional loop for their hatchets. The pikeman's chief weapon had a flat, pointed iron head, an ash pole which varied from 13–18 ft and an iron shoe. At the Restoration all musketeers had a cumbersome matchlock, for which a length of slow match was required, either coiled round the musketeer's belt or carried in his hat until in the presence of the enemy, when it was held between the fingers of his left hand with both ends lighted. The forked wooden rest for the musket, used during the Civil War, was discontinued after 1665. The matchlock gradually gave way to the snaphance or flintlock (see Appendix 4). By 1683 all Guards' musketeers were armed with this weapon. Foot regiments took longer to equip, so that by the end of William III's war with France, about half a battalion still had matchlocks. Grenadiers and fusiliers had lighter flintlocks of carbine bore, known as *fusils*, which were fitted with buff leather slings.

The ammunition for the matchlock, and temporarily for the flintlock, was carried in a collar or bandolier of cartridges over the left shoulder, another relic of the Civil War. This was a leather belt from which hung twelve wooden powder tubes, either painted or leather-covered, a ball bag containing twenty-five bullets, and a priming flask, with the slow match coiled round it. This was a noisy and unwieldy piece of equipment to be burdened with. However, by 1686, a paper cartridge containing bullet and charge was in use. A 'cartouch box' (buff leather pouch suspended from a shoulder belt) had been devised to carry these, and was issued to most regiments by 1693. The date such pouches first appeared is uncertain. The Royal Regiment of Fusiliers (later 7th Foot) had them on formation in 1685. The pouches were ordered for the 1st Guards in 1684, but as the musketeers of both Guards regiments were described by Sandford at James II's Coronation as wearing bandoliers, the grenadiers carrying cartouch boxes, the order may, at first, have only applied to the latter. The aforementioned portrait of a grenadier officer of the 1st Guards, c 1685, shows him with a large pouch, presumably containing grenades, slung from a shoulder belt, and a smaller pouch on the front of the waistbelt which probably contains cartridges. This system will be met later in 1751, but at sometime in the intervening period grenadiers were given larger pouches with separate divisions for cartridges and grenades; the latter were 2–3 inches in diameter.

If it is difficult to be precise over the men's dress in this early period, it is even more difficult in the case of officers. In general, its cut followed the men's but in greatly superior materials. No regulations governed officers' dress, other than a colonel's wishes, and the quality and embellishment of an officer's dress depended upon his finances. Nevertheless, crimson, scarlet or red coats seem, generally, to have been worn — though officers' coats seldom carried the regimental facings used by their men. Sandford describes the variety within one regiment alone, the 1st Guards: 'some in coats of cloth of gold, others in crimson velvet imbroidered or laced with gold or silver; but most of them in fine scarlet cloth, buttoned down the breast and on the facings of the sleeves with silver plate [buttons].' These variations in colour and material may simply be a reflection of individual officers' personal wealth, but it is more likely that they indicated gradation in rank between field officers, captains and subalterns.

Although the quality of the cloth and gold or silver lace readily distinguished an officer from a soldier, and a more elaborate coat distinguished a field officer from a company officer, relative rank could be discerned by the sash, the gorget and the weapon. The sash or scarf, originally worn over the shoulder but, by 1685, worn round the waist, seems to have varied in richness according to rank, as Sandford described: 'Their scarfs were either network or gold or silver, or taffata richly fringed with gold and silver.' The gorget derived from the neck-piece of a suit of armour, and was suspended round the neck by a ribbon. In 1685 captains' gorgets were 'silver plate double gilt', lieutenants' gorgets were 'steel polished and sanguined [black] and studded with nails of gold', and ensigns' gorgets were 'silver plate'. All officers carried swords slung from an embroidered baldrick over the right shoulder but increasingly, towards the end of the century, swords were worn from a waistbelt worn under the coat. In addition, field officers, when dismounted, had a 9 ft half-pike, captains had a pike or 8 ft spontoon, lieutenants carried a partisan up to 1684, thereafter a pike, and ensigns (unless carrying the Colours) carried a half-pike. Grenadier and Fusilier officers carried partisans up to 1684, thereafter a light musket or fusil. The 7–8 ft halbert with axe blade was the distinguishing mark of a sergeant.

The 18th Century

BACKGROUND

Throughout this century the Army was engaged in the long struggle against France, with operations ranging far beyond Europe and the Mediterranean to North America, the West Indies and India: the Wars of the Spanish and Austrian Successions, the Jacobite Rebellions, the Seven Years War, the American Revolution, during which France, Spain and Holland also declared war on Britain, and from 1793 to the end of the century the French Revolutionary War. Though French power in India was crushed by 1763, the East India Company's armies aided by some King's regiments were almost constantly in action, chiefly against Mysore and the Mahratta Confederacy up to the end of this period.

At its greatest strength in the War of the Spanish Succession, the British Infantry totalled seventy-nine battalions but many were disbanded at the Peace of Utrecht, leaving an establishment of three Guards regiments and thirty-nine of Foot. Just before the outbreak of the French Revolutionary War, for which many new regiments were raised, the regiments of Foot numbered seventy-seven. Although regiments continued to be known by their colonels' names up to 1760, their numbered precedence was first promulgated in the 1742 *Army List*, and in the 1754 *Army List* the colonels' names were omitted for the first time. If a more junior, higher-numbered regiment was disbanded after a war, its number was held in abeyance to be re-allotted when new regiments were raised.

During the first half of the eighteenth century the battalion was reduced to nine companies, one being the grenadiers, who remained the picked men of a battalion, though the grenade itself fell into disuse. The Seven Years War witnessed the increasing use of light troops by all armies. The first light troops in the British Army were the Independent Companies raised for policing the Highlands of Scotland. In 1739 these were formed into the first Highland Regiment (later 42nd Foot). A second regiment (Loudon's) and more

Infantry halted during the War of the Spanish Succession. At the right of the group sits a sergeant, identified by his halbert. Painting by Marcellus Laroon.

companies were raised during the 1745 Jacobite Rebellion. These were disbanded in 1748 but many more were raised for the Seven Years War and the American Revolution, though disbanded thereafter. Before the peace-time Army was increased in 1793 it had six Highland regiments (42nd, 71st–75th). After the Seven Years War, Highlanders became normal infantry, but special light infantry battalions had been formed during that war in Canada and in Germany. These were temporary units only, but in 1770 a tenth, light infantry, company was added to each infantry battalion. As this company's post was on the left of the battalion line, and the grenadiers' post was on the right, they were known collectively as 'flank companies'. During the American Revolution it became common practice to group such companies into special battalions for action.

Pikemen were not used in battle after the seventeenth century, and from the War of the Spanish Succession onwards the battalion line was formed in three ranks, with the junior officers and sergeants in a fourth, supernumerary rank behind. Musketry volleys by ranks gave way to platoon fire, which enabled a continuous and better-controlled fire to be kept up. On going into action the battalion was divided into eighteen platoons, told off into three 'firings' each of six platoons. The platoons of each firing were staggered along the line and discharged their volleys simultaneously, the front rank kneeling, the second and third standing, the latter firing between the intervals of the former. Highly effective under Marlborough, this system was refined and improved by Prussian drill methods as the century progressed. Though improvements to the musket, particularly the introduction of steel ramrods, quickened the rate of fire (see Appendix 4), the mean range for effective volleys remained at about 60 yards (55m) throughout the eighteenth century. Much of the reputation won by the British Infantry in Marlborough's victories, at Dettingen and Fontenoy, Quebec and Minden, in America and India was due to the excellence of its musketry.

DRESS 1701–1741

Under Queen Anne the scale of clothing for the infantryman was laid down in 1708 as a 'full-bodied cloth coat, well lined', a waistcoat, a pair of kersey breeches, a pair of 'good strong stockings' and shoes, two shirts and neckcloths, and a hat 'well laced'. In the following year, the infantryman would receive the same again, except that the waistcoat would be made from the previous year's coat. In addition he was issued with a sword, belt and cartridge pouch. Sergeants, corporals and drummers had the same scale 'but everything better of its kind'.

Grenadier of a Scottish regiment of Foot, or possibly the 3rd Guards, c 1715. Note the lace loops common to grenadiers' coats at this period. From a recruiting picture board.

In essentials, therefore, the basic clothing and accoutrements hardly changed from the previous reign. The general appearance of Queen Anne's

Half-pike drill for officers, 1726. The drawings were made by William Hogarth for the Honourable Artillery Company but the dress is typical for Infantry officers of George I's reign.

soldiers can be seen in the tapestries and paintings by Laguerre, commemorating Marlborough's victories, though the paintings may have been based on uniforms of the next reign. Apart from a few figures which appear to represent the Foot Guards, these sources show no specific regiments, and even an obvious red coat does not necessarily make its wearer British. There is, however, some documentary evidence for the period — clothing lists, inspection returns, advertisements for deserters — in which regimental dress distinctions can be found, and on which the figures in Plate 2 are based.

By now, the hats were turned up on all three sides, with the brims usually laced. Grenadier caps still seem to have varied in style, but the cloth front was assuming a more pointed shape and the crown of the bag, instead of being left to hang free, was beginning to be fastened up to the front, giving the mitre shape which would soon be universal. Guards caps seem to have been of this style. The full-bottomed wig remained in fashion for officers but its ends were sometimes plaited together; the men's hair was worn a little shorter and occasionally tied back with a ribbon.

The coat was unbuttoned on the chest, revealing the waistcoat, and the skirts of the coat acquired a fullness from pleats falling from buttons at either hip. Since the front buttons, below the waist down to the hem, were never fastened, they were dispensed with. The facing colours still showed on the deep cuffs, and the coloured lace loops remained the prerogative of grenadiers, though the practice was starting to spread.

Marlborough himself often wore long gaiters, buttoning down the outer leg, in preference to riding boots, and this type of legwear, usually in brown canvas, is visible on infantrymen in representations of his victories. From 1710 he ordered that 'the Foot be provided with white gaiters, both officers and soldiers'.

It is clear from the tapestries commemorating Marlborough's victories that, in addition to the basic accoutrements of cartridge pouch and sword belt, the soldier on service had a knapsack, apparently of canvas and slung over one shoulder, and that other items such as cooking pots and tent-poles were divided up among the men of a company. On going into action, however, such impedimenta were laid aside.

Marlborough's 1710 order required that the soldiers' clothing be uniformly made, 'the colour

Left: Grenadiers of the Princess of Wales's Own Regiment (later 2nd Foot), c 1727. The coat lapels are now buttoned back and the lace is of special regimental pattern. The bayonet hangs from a frog on the left front of the waistbelt and the sword from a separate frog at the left side. On the cap's little flap is the lamb, the special badge of this regiment. From recruiting picture boards.

of the linings, facing and looping excepted'. It also hinted at irregularity in officers' dress, for it stipulated that they 'be all clothed in red, plain and uniform, which is expected they shall wear on all marches and other duties as well as days of Review, and that no officer be on duty without his regimental scarf (sash) and spontoon.' The latter weapon was now to be carried by all officers.

Marlborough's insistence on greater uniformity was continued by the Hanoverian kings, particularly George II who promulgated a number of orders to enforce uniformity and curb the whims of the regimental colonels in the matter of dress. The first was issued in 1727, requiring all regiments to have 'a fixed clothing', each differing in its facings, and that once fixed it should remain so, 'whosoever shall come at the head of the corps'. A change of facing colour now required the Royal Assent, as in 1733 when Kane's Regiment (9th Foot) asked to be allowed to revert from bright green to orange, the colour worn in 1687, out of the colonel's 'respect for the House of Orange'.

Two Hanoverian emblems of long-lasting usage now appeared — the black cockade, adopted in 1715, and attached to the left front of the hat with a button and loop; and the white horse, which in time would be emblazoned on all grenadier caps, and which continued as part of certain regiments' insignia until recently. All grenadier caps now assumed the mitre shape, the bag remaining red, the surrounding turn-up and pointed front (to which the bag was attached with a tuft) being in the facing colour. A little flap turned up in front, embroidered with a motto or title and, in most instances, the white horse. The colonel's crest on the front, though permitted for a while, was replaced under George II by either the Royal cypher or a special badge approved by the King. Hair or wigs were always tied back and in 1739

Plate 2: 1701–1741

4. Drummer, Lucas's Regiment, 1702 (later 34th Foot). **5. Private, Erle's Regiment, 1709** (later 19th Foot). **6. Grenadier, Wynn's Regiment, 1708** (disbanded). These three figures are typical of Marlborough's infantry. The felt hats were beginning to assume the tricorne shape. Drummers adopted the pikemen's custom of wearing reversed colours. Fig. 4, based on a *London Gazette* description, has his colonel's crest emblazoned on the back of his coat and on the drum. The latter is suspended from a belt, covered with facing cloth and edged with lace, slung round the neck. Grenadiers wore caps of various designs. Fig. 6 has an upright front, covered in facing cloth, and the colonel's crest of a wolf's head, also according to the *London Gazette*. The practice of reinforcing buttonholes and coatseams with coloured lace was, at first, confined to grenadiers and drummers. Fig. 5, based on a clothing list, wears canvas gaiters to protect his stockings. The cartouch box has replaced the bandolier, and the bayonet (first the plug type, Fig. 6, then the socket pattern, Fig. 5) is carried in a frog on the front of the waistbelt. The sword hangs from another frog on the left side. A canvas knapsack, slung over one shoulder, contains the soldier's personal campaigning kit.

4. Dmr, Lucas's, 1702.

5. Pte, Erle's, 1709.

6. Grenadier, Wynn's, 1708.

7. Grenadier, Wills's, 1725.

9. Pte, Crawford's, 1740.

8. Offr, R. Welsh Fuziliers, 1740.

officers were to appear powdered at reviews.

The propensity of the coat fronts when unbuttoned at the chest to curl outwards, thus revealing the lining, gradually developed into lapels, which were permanently buttoned back down to the waist, the edges and buttonholes being sewn with lace. This seems to have been customary by the end of the 1720s. The famous picture-board figures of Kirke's (2nd Foot), dated between 1715–1727, have them, as did Disney's (29th), in 1727, in yellow. The latter's coats also had a turned-down cape or collar, about two inches deep, a by no means universal feature but one shown on the 1st Guards coats in Laguerre's Blenheim painting. As the scanty pictorial evidence of the 1715–1740 period only depicts grenadiers, it is difficult to say when the battalion men's coats acquired lacing. A Foot Guards order of 1737 indicates that the practice had begun of looping up the front and back corners of the skirts to disencumber the legs when marching. This revealed more of the breeches and hose, the former being mostly red (blue for Royal regiments), the latter white, though these were frequently covered by the white gaiters.

Orders to the Foot Guards in 1735–1737 show that shoes had square toes, and specify how hats and accoutrements were to be worn. The former were to be 'well-cocked, and worn low so as to cover their foreheads, and raised behind with their hair tucked well under; the point of their hats pointing a little to the left.' The pouches were 'to hang (on the right side), with the foresling

Viscount Southwell, Coldstream Guards, 1739. His sash, gorget, gloves, gaiters and spontoon show that he is dressed for duty. Note the hat under his left arm.

Plate 3: 1701–1741

7. Grenadier, Wills's Regiment, c 1725 (later 3rd Foot). **8. Company Officer, Royal Welsh Fuziliers, 1740** (later 23rd Foot). **9. Private, Crawford's Highland Regiment, 1740** (later 42nd Foot). The first two Hanoverian kings strove to achieve greater uniformity in the Army's dress, and, though the colonels' proprietary rights over their regiments in the first reign is reflected in the personal crest on Fig. 7's cap, the cap has now assumed the familiar mitre shape. The rest of the clothing is little different from that of the previous reign, except for the white gaiters, now a regular feature of the uniform, and a white stock instead of the neckcloth with loose ends. Officers still enjoyed considerable latitude in their dress and Fig. 8, based on a portrait, shows a very plain coat but a richly embroidered waistcoat. Blue breeches, to match the facings of Royal regiments, were becoming customary for the men, though these were not always adopted by the officers. The gorget is suspended by a ribbon in the facing colour, and rank is also shown by the gold aiguillette on the right shoulder. All hats bore the black cockade of Hanover. Fig. 9's dress followed the normal Highland costume but with red jacket and belted plaid in the Government tartan. Each man was issued with musket, bayonet and broadsword, but was also permitted, at his own or his colonel's expense, the other Highland weapons, the dirk, pistol and targe. The dirk and sporran were suspended from a concealed waistbelt which supported the belted plaid under the waistcoat; a separate belt supported the pouch and bayonet.

Grenadier sergeant and drummer of the Foot Guards, c 1740. The sergeant's rank is shown by his sash and halbert. Detail from a painting by Jacques Laguerre.

buckled under the sword-belt, which belt is to be on the outside of the coat, buckled tight, the coat pulled down so as to sit well and even.'

The height for guardsmen was fixed in 1729 as 5 ft 9 in., an inch shorter for other regiments. Sergeants continued to be distinguished by their sashes and halberts and from 1729 corporals wore a white shoulder knot.

There is little evidence as to officers' dress in this period, but portraits reveal that regimental facings on the coats were more usual than in the previous century. The 1737 Guards order states that the officers were to appear in 'their new regimental clothes, gaiters, square-toed shoes, gorgets, sashes, buff-coloured gloves, regimental laced hats, cockades, the button worn to the left side, and twisted wigs according to pattern.'

A major improvement to the infantryman's arms occurred with the issue, from c 1725, of an improved flintlock, the Long-Land pattern (see Appendix 4), generally with brass furniture and with its brown wooden parts polished instead of painted black as was done previously, which contributed to its nickname of 'Brown Bess'. From 1726 the wooden ramrod gave way to a ramrod of steel. For a period following the general introduction of the bayonet in the late seventeenth century, the battalion companies may have dispensed with their swords, but from 1721 all NCOs and men were ordered to wear bayonets and swords, or hangers. In 1729, at Gibraltar, Disney's Regiment was asking for broadswords as 'our hangers are most of them broke' and 'most of the other regiments have broadswords.'

Grenadier of the 1st Guards, 1740. His cap bears the garter star and the white horse. The pouch flap has the Royal cypher, a distinction peculiar to this regiment. The sword and bayonet scabbard now hang from a combined frog at the left side. Note the width of the musket sling. Watercolour by B. Lens.

Towards the end of this period a new type of infantryman appeared, the Highlander. Independent companies of Government Highlanders had existed between 1667 and 1717, probably wearing their own clothes: not until they were re-raised in 1725 is there evidence of any uniformity of costume. Company officers were to provide NCOs and men with 'a plaid clothing and bonnet in the Highland dress, the plaid of each company to be as near as they can of the same sort or colour.' Such uniformity of plaids was a novelty in the Highlands, for, contrary to some beliefs, the notion of a uniform tartan for each clan at this date is a myth. From this ruling of uniformity of plaid for Government Highlanders there appeared the dark blue, green and black sett known as the Government (later Black Watch) tartan. Since the normal coat and waistcoat of the day were quite unsuitable for wear over the voluminous folds of the belted plaid, which was the basic Highland garment, the coats, or rather jackets, and waistcoats of Highland dress were short. When the Highland Regiment was formed in 1739 its upper garments were red, faced buff, to ensure easy recognition of its military status in the Highlands. This may also have been the case for independent companies before this date. The first dress of the Highland Regiment is shown in Fig. 9, Plate 3.

DRESS 1742–1750

In furtherance of George II's aim to regulate the Army's dress, there appeared in 1742 the important '*Representation of the Cloathing of His Majesty's Household and all the forces upon the Establishments of Great Britain and Ireland*' which was followed in 1747 by a '*Regulation for the uniform Cloathing of the Marching Regiments of Foot*', but as the latter's contents were consolidated in the *Royal Clothing Warrant* of 1751, they will be considered under the next period.

Devised by the Duke of Cumberland, the 1742 book illustrated for the first time the dress of every regiment and gave the Royal authority to a style of uniform and to regimental distinctions that must have been in use for some years. For the Infantry it showed a battalion private of the 1st, Coldstream and 3rd Guards, forty-nine regiments of Foot (one of Invalids), the Highland Regiment, and eight Independent Companies; the latter being not the independent companies in the Highlands, but units garrisoning certain colonies across the Atlantic. No grenadiers are shown but the style of their caps is evident from caps worn by figures

Three figures from the 1742 Clothing Book. From left: 1st Guards; 7th Royal Fusiliers; 24th Regiment.

Grenadier caps. From left: Officer's, Royal Regiment of Ireland (later 18th Foot), c 1712. Officer's, 43rd Highland Regiment (later 42nd), 1747. Private's, 49th Regiment (later 48th), 1747. Note the Royal cypher and the white horse of Hanover on the latter two caps.

representing the three regiments of Fusiliers. The facings and lace of each regiment are given in Appendix 1.

Apart from the Fusiliers, all the figures wear the three-cornered hat with the black cockade and edged with white lace; a comparison of the figures appears to show two slightly differing styles of hat, one having the sides more tightly pinched in. During the Flanders campaign from 1742, field signs, usually a sprig of oak leaves, were worn in the hats, a seventeenth century custom of distinguishing friend from foe. A surviving grenadier cap from this period, of the 49th (later 48th) Foot, has the Royal cypher on its buff front and the white horse with motto *Nec Aspera Terrent* on the turned-up red flap in front; insignia which were ordered for most regiments in 1747, although some were entitled to special badges, as will be seen in the next period. Any use of a colonel's arms or crest was now expressly forbidden. Grenadiers were issued with oilskin covers for these caps to protect them in bad weather. To further preserve their hats or caps, all men had forage caps made out of old coats, possibly of the style shown in Fig. 14, Plate 5.

It appears from the Clothing Book drawings that the men's hair was cut short, but an order by the Duke of Cumberland in 1742 forbade this. It was, in fact, tied back with a ribbon at the nape of the neck with the ends tucked under the coat collar; for parades it was lightly powdered. David Morier's painting of the 4th Foot at Culloden shows that the grenadiers' hair was plaited and turned up under the caps.

Coats and waistcoats were worn unbuttoned to the waist, revealing a white stock and the shirt ruffle; and the skirts were looped up back and front to fall at the sides. When the garments were fitted, the coat was to be 'rather short than too long', and the waistcoat was to reach half-way between the waistband and the knee. The latter, being made from old coats, were customarily red, but in the 1742 book the 5th and 39th Foot had green, the 20th and 30th yellow, and the 35th orange. Except for seven regiments, including the Invalids and Highlanders, all regiments had lapels buttoned back but in 1743 lapels were ordered for all. As can be seen in Appendix 1, some regiments had plain white lace, others no lace, and the 6th, 8th, 11th, 17th, 22nd, 31st, 36th and 40th had coloured lace binding the coats but simple white braid loops for the buttonholes. The waistcoats were also bound round with lace. For hot weather stations garrisoned by independent companies the coats were of a lighter cloth called camlet, faced green without lace, and lined with brown linen.

In the Clothing Book drawings the coats are fastened at the waist with the waistbelts buckled

Officer, private and sergeant of Highlanders, c 1745. The officer's rank is denoted by the sash over his left shoulder, the sergeant's by his halbert. The private has drawn the upper portion of the belted plaid over his shoulders, as was done in inclement weather. Various Highland weapons, pistols, dirk and broadsword are shown.

round them, but an order issued in Flanders in 1744 required troops to appear at certain parades with their coats open. When so worn, the waist-belts went under the coat, an arrangement that was increasingly adopted except in marching order, when the coat was closed by buttoning the lapels across with the waistbelt outside. Hogarth shows this method in his 'March of the Guards to Finchley' during the '45 Rebellion (see also Fig. 14, Plate 5).

Breeches were now uniformly red, or blue for Royal regiments, though in the Clothing Book the 2nd Queen's have red. White gaiters covering the knee are shown for all but these were reserved for parades. On service a second pair were issued for marching. In Flanders in 1742–1743 these were brown; in the latter year the Coldstream Guards had black, changing to grey a year later, when this colour was ordered for all the Infantry in Flanders. In May 1745 the Guards brigade in Europe reverted to brown. Shoes had 'a broad round toe, high quarters' and fastened with a buckle.

The waist and pouch belts shown in the Clothing Book are of buff leather, broader than before, and were pipeclayed with a mixture of 1 lb of yellow ochre to 4 lb of whiting. The waistbelt now had a combined frog for bayonet and hanger, and the loose end of the pouch belt had a brush and picker for cleaning the musket lock. The pouches were also of buff leather but with the flaps blackened in the Guards and some other regiments. The 1st Guards had the crowned Royal cypher on their pouch flaps. From 1745 brass match cases secured to the pouch belts were ordered for grenadiers. According to Morier's Culloden painting the latter also had a small black cartouch box on the front of their waistbelts, in addition to the large pouch. As in Marlborough's wars, camp equipage (knapsacks, tin water bottles and cooking pots) was issued on service.

Muskets remained as before and the hangers in the Clothing Book have a guard and simple knuckle-bow, although there is evidence that this type of hilt was being superseded by a basket type, particularly for grenadiers.

Officers, for whom there were still no dress regulations, had gold or silver hat lace, matching that on their coats and their buttons. Grenadier and Fusilier officers had richly-embroidered caps like their men, but these tended to be kept for ceremonial occasions. The grenadier officer in Morier's Culloden painting wears a silver-laced hat in contrast to the caps of his men. There is mention in Cumberland's order books of officers wearing wigs 'of regimental pattern', those of the 1st Guards being 'twisted'. The wigs were tied at the back in a *queue* (pigtail) which, in emulation of the Prussian fashion, often hung well down their backs.

Their coats, now generally with laced lapels, more square-cut than the men's, and cuffs in the facing colour, were worn open, revealing the stock, shirt ruffle and waistcoat, and with the skirts hanging freely. In some regiments the lapel buttons and pocket flaps had gold or silver lace loops, those of the Guards being pointed. The laced waistcoats usually matched the coat or the facings but buff was also popular. Aiguillettes on the right shoulder indicated rank, and when on duty crimson silk sashes were worn over the right shoulder with gorgets matching the lace and suspended by a ribbon in the facing colour. In addition to their regimental coats, officers also had a second undress coat or frock, in blue, which could be worn on less important parades. For a review in 1742 buff gloves were ordered for officers of the Guards, and there is evidence of these being worn by the Foot.

Officer of a Royal regiment, c 1745. The facings are blue, the lace gold. Sash and gorget are worn. Painting by John Wollaston.

Breeches were red or blue, sometimes buff, the former colour often being worn even when the

Plate 4: 1742–1750

10. Grenadier, Ligonier's Regiment, 1745 (later 48th Foot). **11. Colonel, Loudon's Highland Regiment, 1746** (disbanded). **12. Sergeant, Battalion Company, Barrell's or King's Own Regiment, 1746** (4th Foot). This plate and Plate 5 are representative of the War of the Austrian Succession and of the '45 Rebellion, and illustrate George II's efforts to impose Royal authority over the Army's clothing by the issue of the 1742 Clothing Book. The Royal cypher and the white horse of Hanover on the grenadier cap (Fig. 10) are further evidence of the regimental colonels' diminished powers. The grenadier has two pouches: the larger to carry his grenades, the smaller, on his waistbelt to carry his cartridges. A hide knapsack is slung over his right shoulder and he wears grey marching gaiters. A brush and picker are attached to the loose end of the pouchbelt and, for grenadiers, a brass match case is fixed above the buckle. Bayonet and hanger share a frog on the left of the waistbelt. The sergeant (Fig. 12) has silver lace loops, instead of the men's coloured worsted. A crimson sash with stripe in the facing colour and the halbert mark his rank. Fig. 11, based on a portrait, shows an officer of the second Highland regiment to be formed. The coat is similar to other officers' but is cut short to accommodate the voluminous belted plaid.

10. Grenadier, Ligonier's, 1745.

11. Offr, Loudon's, 1746.

12. Sgt, Barrell's, 1746.

13. Pte, Graham's, 1746.

15. Fifer, Foot Guards, 1750.

14. Pnr, Coldstream Guards, 1747.

Mid eighteenth century drum of the 6th Regiment, bearing the antelope, the special badge granted to the regiment as one of the Six Old Corps.

men's breeches were blue. Long white gaiters like the men's, were worn on parade but for the march 'spatterdashes' (a form of gaiter) or boots, usually with black or brown tops, were ordered in 1744.

Officers' swords hung from a waistbelt concealed under the waistcoat. The hilts normally matched the coat lace and had a crimson and gold sword knot. Their chief weapon, and a symbol of an officer being on duty, was still the half-pike or spontoon. The latter, which had a broader spear point and a cross-bar, seems to have been generally ordered from 1743–1744. Grenadier officers carried fusils with slings and bayonets fixed.

Sergeants' coats were like the men's coats but of better quality and possibly laced silver (see Fig. 12, Plate 4). Their rank was indicated by their halberts, and by sashes with a stripe in the facing colour which, from 1745, were ordered by Cumberland to be worn round the waist.

Morier's Culloden painting shows drummers of the 4th Foot in low mitre caps with special drummers' insignia, red coats faced blue as befitted a Royal regiment, with lace of the same colour but of a different pattern from the rest of the regiment. Drummers' clothing was not specified in the 1742 book, but the 1747 regulations stipulated the Royal livery (red and blue) for Royal regiments and reversed coats for the remainder.

The Highland Regiment is shown in the 1742 book with bonnet, short jacket and belted plaid as described earlier, though the plaid is of a most curious diamond pattern, probably due to the artist's ignorance of Highland costume. In the bonnet is a short piece of white lace with a red zig-zag, but elsewhere the black cockade is depicted. During the '45 Rebellion many Government Highlanders wore a red saltire pinned to the cockade as an identification mark in contrast to the white cockade adopted by the Jacobites. From a surviving officer's cap, it is clear that, until 1747, the Highland Regiment's grenadier

Plate 5: 1742–1750

13. Battalion Private, Graham's Regiment, 1746 (11th Foot). **14. Pioneer, Coldstream Guards, 1747.** **15. Fifer, Foot Guards, 1750.** The amount and arrangement of coat lace authorised by the 1742 Clothing Book varied between regiments. Fig. 13 has plain white loops round the lapel buttonholes. The pouches of all regiments were of buff leather, matching the belts, but some were blackened, as here. The coat is fastened, as shown in the 1742 book, and the red breeches are now regulation for non-Royal regiments. From c 1725 the Long Land musket (Brown Bess) replaced the old blackened flintlocks. The dress of the Foot Guards was similar to that of the Line, but with plain white lace in all three regiments. Every regiment had a number of pioneers equipped with spades, picks and axes as well as muskets. Fig. 14 wears a cap, which is more suited to pioneering work than the hat, with a badge of crossed tools to signify his trade. His coat lapels are buttoned across, as became customary in marching order when the waistbelt began to be worn under the open coat for normal parades. Drummers and fifers of Guards (Fig. 15) and Royal regiments wore red coats faced blue with special Royal lace; other drummers' coats were in reversed colours. A special feature of their coats was the false sleeves hanging from the shoulders. All had caps similar to grenadiers' caps but lower, with appropriate insignia. Fifes were re-introduced from 1748 and were carried in cylindrical cases.

company wore mitre caps like other regiments. Thereafter bearskin caps with Royal cypher and crown on a red ground on the little flap were allowed for both Highland regiments. An officer of the second regiment, Lord Loudon's is shown in Fig. 11, Plate 4.

DRESS 1751–1767

The style of uniform continued as before during this period, but the waistbelt was no longer buckled outside the coat, except when the lapels were buttoned across in marching order as already described. It is discussed separately because of two important pieces of evidence: the Clothing Warrant of 1751, which embodied the 1747 regulation, and the paintings executed at the same time by Morier of grenadiers of each regiment of Guards and Foot. The facings listed in the warrant are virtually the same as in the 1742 book, but the paintings disclose that many of the lace patterns had become more elaborate (see Appendix 1).

Hats were unchanged, but towards the close of the period the front points, always worn over the left eye, were rather more cocked up. The oak-leaf field-signs re-appeared during the Seven Years War campaigns in Germany. The warrant of 1751 confirmed the 1747 regulation which had stipulated that the front and rear turn-up of a grenadier cap were to be in the facing colour, with the regiment's number in the middle of the latter; the bag red, and the little flap in front also red with the white horse and the motto *Nec Aspera Terrent*. All were to bear the Royal cypher and crown on the front, except for the Royal

Left: Grenadiers of the Coldstream Guards in Flanders, 1747. The coats are worn open, with the waistbelts underneath. The gaiters have been removed. Note the brass matchcase and the brush and picker on the pouch belt. Watercolour by Paul Sandby.

Plate 6: 1751–1767

16. Battalion Private, 47th Regiment, c 1760 (Canada). 17. Officer, 40th Regiment, attached to the Light Infantry, c 1758 (Canada). 18. Officer, 42nd Royal Highland Regiment, 1762. The provisions of the 1751 Clothing Warrant, and the Seven Years War period, are illustrated in this plate and Plate 7. Fig. 16 is in marching order, fully accoutred with hide knapsack, canvas haversack and tin water bottle. Black gaiters with leather tops are worn. Fig. 17, based on an officer's portrait by J. S. Copley, shows how the regulation hat and coat were cut down and modified, for the Light Infantry formed during the North American campaigns. He carries a fusil with short bayonet, powder horn, a cartridge pouch conveniently placed on the waistbelt and a hatchet. His shortened coat has been stripped of all lace and his boots are non-regulation. The Highland officer, Fig. 18, also based on a portrait, still wears the belted plaid and shortened coat, which now has a turned-down collar. The practice of adorning the bonnets with feathers has begun. He is armed with broadsword, fusil and bayonet for which a small pouch is worn on the waistbelt. Curiously no sporran or dirk are shown in the original portrait. His rank is indicated by the sash and aiguillette on the right shoulder. The buttonholes are laced in bastion-shaped loops, a design repeated on the waistcoat.

16. Pte, 47th Regt, c 1760.

18. Offr, 42nd Hldrs, 1762.

17. Offr, 40th att. Light Infantry, 1758.

19. Offr, 21st Fusiliers, 1755.

21. Offr, 84th Regt, 1761.

20. Cpl, 3rd Guards, 1762.

Plate 7: 1751–1767

19. Company Officer, 21st Royal North British Fusiliers, c 1755. 20. Battalion Corporal, 3rd Foot Guards, 1762 (Germany). 21. Field Officer, 84th Regiment, 1761 (India) (disbanded 1764). All ranks of Fusilier regiments wore mitre caps, although officers (Fig. 19) usually retained their richly-embroidered, mitre caps for ceremonial purposes. This figure, based on a Gainsborough portrait, has red breeches rather than the blue worn by the men of Royal regiments. Being on duty he wears his gorget and carries a spontoon which replaced the half-pike as the company officer's weapon from 1743. In Fig. 20 the white shoulder knot indicates corporal's rank. By the 1760s the front of the hat was slightly more cocked up and a field sign of green leaves was worn when serving with the Allied army in Germany. The coat is worn open over the waistbelt and the white gaiters have been blackened, as ordered for the Guards brigade in Germany in 1761–1762. The lace loops of the Coldstream and 3rd Guards had pointed ends, the 1st Guards had square ends. Fig. 21, based on a portrait of Eyre Coote, shows a uniform worn in India. The riding boots are of light and supple leather and the buff waistcoat may have been of a material more suited to the climate. White knee pieces protect the breeches from the boot blacking. The sword is suspended from a waistbelt under the waistcoat.

Left: Grenadiers of the 1st, Coldstream and 3rd Guards, 1751. The right-hand figure and the seated battalion man at the extreme left are both corporals, as can be seen from the white knots at their right shoulders. Though the three uniforms are superficially the same, there are differences in the arrangement of the lace on coats and waistcoats. Painting by David Morier.

regiments: 1st Royals, 2nd Queen's, 4th King's Own, 7th Royal Fusiliers, 18th Royal Irish, 21st Royal North British Fusiliers, 23rd Royal Welch Fusiliers; and what were known as the Six Old Corps: 3rd Buffs, 5th, 6th, 8th King's, 27th Inniskilling and 41st Invalids. These thirteen regiments were permitted special devices on their caps as well as on their Colours and drums. The 1st had the King's cypher within the circle of St Andrew, crowned; the 2nd, 4th, 7th, 8th and 41st all had a crowned garter encircling respectively the Queen's cypher, the King's cypher, the rose, the white horse, and the rose and thistle conjoined; the 3rd had the green dragon, the 5th had St George and the dragon, 6th had the antelope; the 18th had the harp and crown, 21st had the thistle within the motto of St Andrew crowned, 23rd had the Prince of Wales's feath-

Grenadiers, 16th and 17th Regiments, drummer (*back view*) and grenadier, 18th Royal Irish Regiment, 1751. The caps of the first two have the Royal cypher, but the 18th, as a Royal regiment, has its special badge of the harp. The drummer's coat is red, faced blue, laced yellow with a blue stripe, and has hanging sleeves. Painting by David Morier.

Grenadier, 48th Regiment, equipped for service, 1751. Over his right shoulder is a brown hide knapsack and tin water bottle, over his left a grey canvas haversack. Detail of a painting by David Morier.

ers, and 27th had the castle of Inniskilling. The Foot Guards were not covered by the warrant but their caps followed the same pattern, with the garter star and crown as badge.

Except for the 41st Invalids, all the coats in the Morier paintings have lapels and lace loops, the lace loops being either square-cut or pointed. Plain white lace was worn by the Foot Guards, the 1st, 5th and 33rd Foot. A feature of some grenadiers' coats was the semi-circular pieces of laced cloth sewn to the shoulder seams, known as wings. These became regulation for all grenadiers in 1752. In several regiments the waistcoats had loops at the button-holes as well as lace all round.

Breeches, hose, parade and marching gaiters were as before. Up to about 1760 marching gaiters continued to be grey or brown. However, there are references to the Guards brigade and the 12th Foot in Germany in 1761 having blackened gaiters, and an inspection return for the 23rd Royal Welch in 1763 observes that the regiment had 'black gaiters but no white as they never used them in Germany'. A group painting by Morier of c 1763 includes a battalion private, possibly of the 39th Foot, wearing buff or perhaps white breeches with black gaiters, less high than the white variety and cut away behind the knee. The knee pieces may be of black leather as these are clearly shown on a private of a Royal regiment in a painting by Penney of the Marquis of Granby, the British commander in Germany from 1760. Black gaiters seem to have become increasingly common until they were made regulation in the next period, white being retained only by the Foot Guards for parade purposes.

In 1759 regiments destined for tropical climates were to have their coats lined with linen, breeches of the same, white thread or linen hose, and a bladder between the crown of the hat and its lining.

The waistbelt, now worn under the coat except in marching order, was narrower than in 1742, and all grenadiers had the small black cartouch box, mentioned in the previous period for the 4th Foot, strapped round it. The battalion man in the Morier 1763 group has a similar pouch on a black strap, as ordered by Granby in 1762 when the large pouches were modified. In the 1751 paintings all the large pouches are shown as black. The same paintings show well the camp equipage; animal-hide knapsacks and rectangular tin water bottles over the right shoulder, grey canvas haversacks over the left. An order issued in North America, after the opening of the Seven Years War, forbade the old custom of tying tent poles to the muskets, but unfortunately does not say how they were to be carried.

The main change regarding arms was the tendency, contrary to regulations, to dispense with swords for the rank and file (corporals and privates) of battalion companies. The spontoon was fixed as the officer's weapon, except for the fusils of grenadier officers. The 23rd Fusiliers' officers had fusils in 1763, having lost or broken their spontoons in Germany, but they were not authorised for all Fusilier officers until 1770.

No regulations affecting officers' or sergeants' dress appeared in the 1751 warrant but pictorial evidence indicates little change from the previous period. The section dealing with drummers (which included fifers, their instrument having been re-introduced around 1748) regularised what had been common practice for many years: drummers of Royal regiments in Royal livery — red, lined, faced and lapelled with blue and laced with a Royal lace, usually blue and yellow or blue and white; others with coats in the facing colour, lined, faced and lapelled with red and laced at the colonel's discretion, but using the same colours as the soldiers' lace. A feature of these coats was the false hanging sleeves, though it is not known whether all regiments had them. No mention is made of their caps but all wore the mitre type, lower than the grenadiers', with the little red flap bearing the white horse and motto, and trophies of drums, arms and Colours on the front, and a drum at the back on the turn-up. Drums were wood and slung round the neck by a belt, usually laced. The front was painted in the facing colour with, for the Royal regiments and the Six Old Corps, their special devices and regimental number underneath; for the remainder, the King's cypher and crown with the number below.

The 1751 paintings include a grenadier of the 42nd Highlanders wearing the bearskin cap authorised for such regiments in 1747. He has belted plaid and diced hose as before, but his short jacket, though without lapels, has a laced turned-over collar with rounded corners in the facing colour; it is sewn on the outside of the neck opening. All the buttons and holes of the jacket have lace loops. The slashed cuffs and pocket flaps are laced like the normal coats, as is the short waistcoat. The coat is worn open, with a waistbelt supporting a cartouch box bearing the crowned cypher, and a bayonet in a frog. He does not carry the large pouch and shoulder belt, but his broadsword hangs from another belt over his right shoulder. All accoutrements are in black leather. His plain leather sporran (purse with tasselled thongs) hangs from a separate narrower belt. The Government tartan has a red stripe, thought to be a distinction of the grenadier company. Battalion companies wore the flat blue bonnet, as before.

Lieutenant-Colonel Francis Smith, 10th Regiment, 1764. The facings are brownish-yellow, the lace gold. He carries a spontoon. Painting by Francis Cotes.

Ten additional Highland regiments were raised during the Seven Years War, of which two, 87th Keith's and 88th Campbell's, distinguished themselves as light troops in the German campaigns of 1760–1762. Their costume is shown, in some rather crude German engravings, as being similar to the 42nd's 1742 dress, with buff or yellow facings.

In North America similar services were rendered by the 42nd, 77th and 78th Highlanders. The order books of the 42nd show that the belted plaid was often set aside in favour of canvas breeches, blue leggings, or the 'little kilt', which was merely the lower portion of the plaid, and occasionally the two latter combined.

The nature of the terrain and the fighting, particularly against the American Indian allies of the French, underlined the need for more light troops than could be provided by Highlanders alone. Foremost among the protagonists of this new light infantry were the Howe brothers, one commanding the 55th, the other one major of the 58th. Earl Howe, the elder, quickly perceived the impracticality of the regulation clothing, and the force of which he was the second-in-command and leading spirit in 1758 was ordered 'to cut the Brims of their Hats off; no Person, Officer or Private, be allowed to carry more than one Blanket and a Bearskin [rug], no Sash or Sword, nor even Lace . . . the Regulars have left off their proper Regimentals, that is, they have cut their Coats so as scarcely to reach their Waist . . . every Officer to carry his own Pack, Provisions.' Lord Howe was killed at Ticonderoga but his brother, the Hon William, commanded the battalion of Light Infantry during Wolfe's operations against Quebec. Drawn from men of Foot regiments who were 'accustomed to the woods, good marksmen, and alert, spirited soldiers', the Light Infantry received lighter muskets, tomahawks instead of bayonets and special clothing. Hats were converted into caps with upswept peaks and a flap to protect the ears and back of the neck which could be hooked up when not required. The coat retained its lapels but was shorn of all lace, fitted with wings and its sleeves put on the waistcoat, so that it could be dispensed with if necessary. Two extra leather pockets were sewn on the breast to hold balls and flints. The tomahawk hung from a case attached to the waistbelt, a cartouch box on the left side from a shoulder strap, powder horn on the right. The knapsack was carried in a new way, copied from the Indians, which would soon become universal: suspended high on the back by a strap over each shoulder, with the water bottle hanging below. A Light Infantry officer is shown in Fig. 17, Plate 6.

DRESS 1768–1799

The new Clothing Warrant of 1768 was not only more comprehensive, but instituted significant changes to the uniform, some of which had been developing in the previous decade. An illustrated publication on the lines of the 1742 book, but showing a grenadier of every regiment, was produced at the same time. The number of regiments had increased, with new facings to be recorded, and the lace patterns were all simplified (see Appendix 1). Plain lace was now reserved for the Foot Guards and all sergeants. Over the next thirty years modifications to the warrant would be ordered, so that by the end of the century, clothing had acquired quite a different aspect.

Grenadier, 58th Regiment from *Uniforms of the Infantry according to the King's Regulations of the 19th December 1768*. The same basic figure served for each regiment in this document which illustrated the 1768 Clothing Warrant.

Officer of the 31st Regiment, c 1780. The spurs suggest he may be of field rank.

Charles Watson, Grenadier officer of the 25th Regiment, c 1780, dressed according to the 1768 regulations. Note the gorget, sash and twin epaulettes, as specified for Grenadier officers also the stiff-topped gaiters. Painting by David Allen.

Plate 8: 1768–1799

22. Light Infantry Private, 6th Regiment, c 1775. 23. Grenadier Sergeant, 95th Regiment, 1781 (disbanded 1783). **24. Private, Battalion of Light Infantry, 1777 (America).** Some of the 1768 Clothing Warrant changes are shown here: more closely-fitting coats, turned-down collars, narrower lapels, round cuffs, turned back skirts and white waistcoats and breeches. Fig. 23 shows this coat and the bearskin cap with metal plate. His rank is indicated by plain white lace and the sash with facing stripe worn under the coat. The Light companies added to each regiment in 1770 wore shorter coats and special leather caps encircled by chains as in Fig. 22, taken from a painting by De Loutherbourg. The tan belts, powder horn and hatchet were peculiar to these companies. Black half-gaiters, originally confined to Light Infantry, were adopted by all companies on service in preference to the long variety ordered in 1768. The battalions of Light Infantry formed during the American War were dressed more simply; hat turned up on one side only, plain jacket, without facings, buttoning to the waist, possibly made from a cut-down coat, white linen gaiter-trousers and blackened belts (Fig. 24). During this period the bayonet belt moved to the right shoulder and the knapsack was slung high on the back from two shoulder straps. Grenadiers' hair was plaited, tied with a bow and turned up under the caps; the remainder wore theirs clubbed.

22. Pte, 6th Regt, 1775.

24. Pte, Light Infantry, 1777.

23. Sgt, 95th Regt, 1781.

25. Offr, 42nd Hldrs, 1784.

27. Sgt-Maj, De Meuron's, 1799.

26. Pte, 36th Regt, 1791.

Troops of the Minorca garrison, 1770. From left: Grenadiers, 67th, 11th, 13th and 3rd Regiments. The remainder are all officers and men of the 25th Regiment. The clothing and accoutrements, as well as the cap of the 13th, conform to the 1768 warrant. The other caps are non-regulation and the 3rd's type is obsolete.

The Guards were not included in the 1768 warrant but the changes ordered for them over the following six years broadly conformed with the changes ordered for the Foot.

During this period the front cock of the hat gradually became less pointed and more vertical, so that by the 1790s it was developing into a tall bicorne with the left point cocked up and the right point down and slightly forward, often with a feather above the cockade. In 1796 the white binding lace round the brim was changed to black, the feather to be white, and white tufts mixed with the facing colour to be set at each corner.

Grenadier caps changed completely, from cloth to black bearskin, with a black metal plate in front bearing the Royal crest and *Nec Aspera Terrent* in silver. The back part was red, with the special badges of Royal regiments and the Six Old Corps embroidered in white, or with a grenade and regimental number for the remainder. Similar caps were ordered for Fusiliers, though less high and without the grenade. Grenadiers of the Guards also had fur caps but with different plates: 1st Guards had white on black metal, Coldstream had white on red, the 3rd had white on white.

The Light Infantry companies added to

Plate 9: 1768–1799

25. Lieutenant-Colonel, 2nd Battalion, 42nd Royal Highland Regiment, 1784 (India). 26. Grenadier Private, 36th (Herefordshire) Regiment, 1791 (India). 27. Sergeant-Major, De Meuron's (Swiss) Regiment, 1799 (India). These figures are based on contemporary pictures showing the dress worn during the Mysore campaigns. The Highland bonnet (Fig. 25) has become more upright, with a diced band, and the ostrich feathers are more extensive. The coat in the original portrait has no buttons on the lapels, which are probably false, suggesting it is a light-weight pattern designed for the climate. Linen waistcoat, breeches and stockings are worn in preference to the belted plaid, and a lighter weapon than the broadsword is suspended from a belt under the waistcoat. Fringed epaulettes indicate rank. Instead of the bearskin cap, Fig. 26 wears a hat, now cocked quite differently, with a feather of regimental pattern. A regulation coat is worn over a linen waistcoat and gaiter-trousers. This regiment's grenadiers wore moustaches. By the end of the century, the soldier's coat had lost its lapels and was buttoned down to the waist, with a standing collar (Fig. 27). The round hat, with or without fur crest, was frequently worn in hot climates in the 1790s. White pantaloons and half-gaiters replace breeches and knee-length gaiters. The Sergeant-Major, as a regiment's senior NCO, has silver lace and an epaulette on the right shoulder. De Meuron's was one of several foreign regiments in the Army at this time.

Officer of the Light Infantry Company, 87th Regiment, c 1783. He has fringed wings on both shoulders. Painting attributed to Beechey.

Hugh Montgomerie, 12th Earl of Eglington, in the uniform of the 77th Highlanders, c 1780. The facings are green, the lace silver and the tartan Government. Painting by J. S. Copley.

Plate 10: 1768–1799

28. Company Officer, 90th (Perthshire Volunteers) Regiment, 1794. 29. Battalion Private, 3rd Foot Guards, 1799. 30. Battalion Officer, 56th (East Essex) Regiment, 1799. This plate illustrates the British regiments in Europe during the French Revolutionary War. The 90th (Fig. 28) was dressed and trained as Light Infantry from its formation, though not formally so designated until 1815. Light troops now wore a variety of headdress, the Light Dragoon helmet shown here being one of the most popular. The jacket had wings under the epaulettes and short skirts; the red waistcoat was peculiar to Light Infantry. Buff-faced regiments' nether garments and belts matched their facings, as can be seen in the pantaloons. Figs. 29 and 30 show how the hat had assumed a bicorne shape, and how the coat had developed for officers and men after the cut-away style was abolished in 1796; the men's having short skirts. White overall trousers were worn to protect the breeches on service, when haversacks and water bottles were also issued. Officers' coats still had lapels in the facing colour, but in this plate they are shown buttoned across. Fig. 30 has the straight 1796 sword instead of the curved sabre affected by Light Infantry and later Grenadiers. Flank company officers wore epaulettes on both shoulders, battalion officers only on the right. Flank companies still wore their hair plaited and turned up but the remainder's hair was now queued with an 11 inch pigtail.

28. Offr, 90th Regt, 1794.

30. Offr, 56th Regt, 1799.

29. Pte, 3rd Guards, 1799.

Battalion officer and sergeant, 1st Guards, 1792. By this date the coats had standing collars. Engraving after Edward Dayes.

Drummer and battalion private, Coldstream Guards, 1792. Engraving after Edward Dayes.

Grenadier sergeant and private, 3rd Guards, 1792. The buttons of the three Guards regiments are now arranged regularly, in pairs or threes. Note the knapsack beside the private. Engraving after Edward Dayes.

battalions in 1770 generally had small leather caps — see Fig. 22, Plate 8 — but some regiments devised their own headgear and there were several varieties. By the end of the century the most common type was the Light Dragoon helmet with fur crest — see Fig. 28, Plate 10. The Guards had no light companies until 1793, when a round hat with fur crest was adopted. Similar hats, with or without the crest, and sometimes in white, were used in hot climates by all types of infantry.

From 1776, the hair of all ranks was tied in the style known as clubbed. Powdering continued until abolished in 1795, and in the following year hair was to be queued, with a pigtail 11 inches long. Grenadiers, however, continued to wear their hair plaited and tied with a ribbon, the end being turned up under the cap with a comb — a style also ordered for the Light Infantry.

The coats now became more close-fitting than before, with a turned-down collar, lapels 3 in. wide down to the waist, and neat round cuffs $3\frac{1}{2}$ in. deep, all in the facing colour. White metal buttons (with square or bastion-shaped lace loops) fastened the collar, lapels, cuffs and pockets. The skirts, less full than before and shorter for Light Infantry, had white linings and were permanently hooked back. Laced and fringed wings for grenadiers and subsequently Light Infantry, were blue for the Guards, red for the Foot. The shoulder straps, at first red, were to be in the facing colour from 1784. In 1796 the collars were made to stand up, as had been the practice for some years, and the lapels, which hitherto had curved away to reveal the waistcoat, were to be

Officer and private, 7th Royal Fusiliers, 1792. The fusil carried by the officer was dispensed with after this date. Engraving after Edward Dayes.

closed by hooks and eyes to the bottom of the lapel, thus concealing the waistcoat. The following year the lapels were removed, so that the coat now became a single-breasted garment, buttoning down the front to the waist, with lace loops on either side of the fastening. The closing of the coat front automatically shortened the skirts.

From 1768 the coloured laced waistcoats and breeches changed to plain white, or buff for regiments faced with that colour. Light company waistcoats were usually red. Black linen, later woollen, gaiters with stiff tops became universal, except for the Foot Guards on ceremonial occasions. The stiff tops were removed in 1784. Half-gaiters which came up to the calf were ordered specifically for Light Infantry in 1771, but were also worn by other companies, particularly on service. Alternative nether garments were white gaiter-trousers, made in one piece and worn in America, India and by the Guards light companies in 1793. From 1790 white trousers began to appear, either for wear in hot climates or as overalls to protect the breeches.

The belts were now narrower, $2\frac{3}{4}$ in. wide for the shoulder belt which suspended the pouch, 2 in. for the waistbelt. Both belts were pipeclayed white or buff to match the waistcoats. Light companies had belts of tan leather. In 1768 swords were finally abolished for the rank and file of battalion companies. The waistbelt, now supporting only the bayonet, continued to be worn under the coat but, as time went by, men began to wear the belt over the right shoulder, so that it became a second cross-belt. This, apparently, was formally approved in 1784 when the width of both belts was equalised at 2 in., later $2\frac{1}{2}$ in. The Guards' belts were fractionally wider, and the ends of the pouch belt passed through two rectangular brass buckles, each attached to short separate straps which, in turn, were secured to small buckles on the underside of the pouch; the Foot's pouch belts fastened directly on to the latter. The pouch on the right side held thirty-two rounds but in 1784 an additional magazine containing a further twenty-four rounds was authorised. This was to be attached to the bayonet belt when required. In due course the plain buckle of the bayonet belt developed into the shoulder belt plate, which secured the two belts where they crossed on the chest, and also afforded a further ornament on which the regimental number and distinctions could be displayed. The animal-hide knapsack, now always slung over both shoulders, gave way to a folding type with linen pockets, covered in painted canvas. Towards the end of the century a new water container appeared: a circular keg-type of blue-painted wood slung from a brown leather strap. New muskets came into service in this period (see Appendix 4) and in 1784 the grenadiers' swords were abolished.

The 1768 warrant was the first official document to regulate officers' dress. The general cut conformed with the men's, but with gold or silver lace as specified for each regiment on the hats (see Appendix 1) and, if a colonel so desired, on the coats. The buttons matched the lace. Waistcoats were to be plain. Epaulettes, in gold or silver lace or embroidery with a fringe, made their first appearance as an indication of rank, worn on the right shoulder by battalion company officers, on both by Grenadier and later Light Infantry officers. The crimson silk sashes were to be worn round the waist under the coat. Gorgets matched the lace and were engraved with the King's arms, the regimental number and any special badge to which the regiment was entitled. From 1796 they were to be gilt for all, with the Royal cypher and crown engraved; the suspending ribbon in the

Battalion officer and private, 8th or King's Regiment, 1792. Engraving after Edward Dayes.

Private, 42nd Highlanders, in the Low Countries, 1793. The lapels, though unclear in reproduction are correctly drawn, with bastion loops. Drawing by an unknown Dutch or Belgian artist.

Private, Light company, Coldstream Guards, 1797. Guards battalions did not receive Light companies until 1793. Note the gaiter-trousers and round hat with fur crest and green feather. Engraving after Edmund Scott.

facing colour was to be red for regiments faced black.

Sword hilts also matched the lace but the crimson and gold knots were common to all. The spontoon continued as the battalion company officer's weapon until abolished in 1786. From 1788, officers on duty were to sling their swords over their right shoulders outside the coat, off duty under the coat. A regulation-pattern sword was introduced for the first time in 1786, being superseded by a new pattern in 1796. In the 1768 warrant Grenadier officers were to wear bearskin caps and to have fusils, shoulder belts and pouches. In Copley's painting of the French attack on Jersey in 1781, a Grenadier officer of the 95th Regiment is so equipped but wears a hat with a white feather, suggesting that such officers kept their caps for parade wear, as in former times.

The 1768 provision also applied to Fusilier officers from 1770, but was cancelled for both the Grenadiers and Fusiliers in 1792, except for the caps. The abolition of coat lapels in 1797 did not apply to officers, whose uniform henceforth developed marked differences in cut from the men's and continued to do so for nearly sixty years. The coat could be worn in one of three ways: with the lapels buttoned back to show the facing colour and fastened to the waist by hooks and eyes; with the lapels buttoned across, making the coat double-breasted and concealing the facing colour; or in the latter fashion but with the tops of the lapels unbuttoned and turned back to show part of their colour.

Sergeants' hats had silver lace and their coats followed the men's style but with plain white lace. Crimson worsted sashes with a stripe in the facing colour were worn round the waist by sergeants; if the facings were red, the stripe was white. Halberts were replaced by 9 ft pikes in 1792 for all sergeants, including Grenadiers and Fusiliers, who formerly had been armed like their officers; only Light Infantry sergeants retained the fusil. In 1768 corporals' coats received a white silk epaulette on the right shoulder instead of their shoulder knots.

Bearskin caps were ordered for drummers in the warrant and their coat colours remained as before except that, in regiments faced red, the coats were to be white, lined and faced red. Hanging sleeves were abolished. All regiments faced red, buff or white were to wear red waistcoats and breeches. In 1796, regiments faced black were to have white coats faced black, white breeches and waistcoats. All drummers and fifers had a short sword with scimitar blade.

Pioneers were to be equipped with an axe, a saw and an apron. Their caps had a leather crown, bearskin front and red-painted metal plate bearing the King's crest plus an axe and saw in white; the regimental number was placed on the back of the cap.

When the 1768 warrant was issued, the 42nd was once again the only Highland regiment, although more would follow during the wars of this period, including four specially raised for service in India in 1787, who received white hats and probably did not wear the Highland dress on the sub-continent. Apart from grenadier caps, the normal Highland headdress remained the blue bonnet. This began to be more blocked out, with a diced band; the practice of attaching feathers or bearskin tufts behind the cockade increased as the period progressed. Coats and waistcoats conformed to the normal infantry pattern, but were cut short to accommodate the belted plaid. The latter was discarded by the 42nd in America in 1784 in favour of white ticken trousers with black half-gaiters. The appearance of the 42nd was further assimilated to the rest of the Foot, in 1791, when white accoutrements replaced black. Broadswords and pistols for the rank and file were dispensed with in America and were not resumed thereafter.

The 19th Century

BACKGROUND

The Infantry was substantially increased during the Napoleonic Wars, not only by the formation of new regiments (up to 103rd Foot), but also by raising second, and in some cases third and fourth battalions, in a large number of existing regiments. In 1809 the 1st Guards had three battalions, the Coldstream and 3rd Guards two each. The 1st Foot had four, the 14th, 27th and 95th three each, the 60th no less than seven, and sixty-one other regiments two each; the remaining thirty-seven only one.

After Waterloo all the Line regiments reverted to single battalions, except for the 1st and 60th each with two. The 95th, taken out of the Line in 1816 and renamed the Rifle Brigade, also had two. Regiments above the 93rd were disbanded.

Though peace prevailed for Britain in Europe, other than the Crimean War of 1854–1856, the Army was faced in this century with a series of almost continual colonial campaigns all over the world, beginning with the Gurkha War of 1816 and culminating in the South African War of 1899. Operations occurred as far afield as Canada, China and New Zealand but predominantly in India and Africa.

The problems of garrisoning the Empire led to the raising in 1823–1824 of new regiments numbered 94th–99th, numbers previously borne by other regiments. The 60th, entirely a Rifle Corps since 1824, and the Rifle Brigade acquired third and fourth battalions in 1855 and 1857 respectively. In 1858 the 2nd–25th Foot all received second battalions and the 100th, raised in Canada, was added. Two years later nine European regiments of the former East India Company's armies were transferred to the Line as the 101st–109th Foot. Of these, the 103rd or Royal Bombay Fusiliers could claim as ancient a lineage as any of the Line, being the descendant of that regiment raised by Charles II to garrison Bombay; the 101st and 102nd both dated their formation from the mid eighteenth century.

In 1881 the numbered regiments of the Line became territorial regiments, each of two Regular battalions. This required all other than the 1st–25th and the 60th to be amalgamated in pairs (see Appendix 3). From 1896 the Coldstream and Scots Guards acquired third battalions and certain regiments recruiting from centres of large population raised third and fourth battalions.

The French practice in the Revolutionary and Napoleonic Wars of masking the movement of infantry columns with swarms of skirmishers revealed the need for more British light infantry

6th Regiment, c 1802. Grenadier and battalion officers. Though naive in style, this and the following three plates by an unknown hand, possibly a member of the regiment, are an accurate and detailed rendering of the uniform at the beginning of the nineteenth century.

6th Regiment, c 1802. Drum-major and bandsman. Coats yellow, faced red. Lace, (*left*) white or silver, (*right*) red and white. Note the antelope badge on the right arm.

than was provided by the battalion light companies. One Light Infantry and two Rifle battalions were added to the 60th in 1797–1799; the Rifle Corps, later 95th, was formed in 1800; and between 1803–1815 the 43rd, 51st, 52nd, 68th, 71st, 85th and 90th were converted into Light Infantry regiments. Besides outpost duties, the task of light troops and riflemen was to inflict maximum damage on the enemy, both in attack and defence, by driving off his skirmishers and weakening him before he was faced by the main battle-line of the ordinary infantry, now reduced to two ranks deep to give maximum frontage for its musketry volleys.

As the century progressed and colonial wars called for different tactical techniques, all companies of a battalion learned the skirmishing drills of the Light regiments and companies, and independent firing became an alternative to volleys. In time, therefore, the designation *Light Infantry* acquired an honorary, rather than a tactical significance, as had *Fusiliers* in former days, and as such was awarded for distinguished service, e.g. to the 13th in 1822 and the 32nd after the Indian Mutiny. The 5th and 87th became Fusiliers in 1836 and 1827 respectively. In 1860 the titles and dress distinctions of the Grenadier and Light Infantry companies were abolished and a battalion's companies were numbered consecutively throughout.

With the advent of the long range breech-loading rifle in the last quarter of the century, the battlefield tactics of close-order line and column, protected by skirmishers, which had served in the

6th Regiment, 1802. Grenadier corporal and Light Infantry sergeant. The grenadier's hair is plaited and turned up under the cap, his forage cap is rolled on top of his pouch.

6th Regiment, c 1802. Light infantryman and battalion private in undress. Note the length of the coats at this date.

Crimean as well as in the Peninsular War, were no longer appropriate. More emphasis was placed on extended order, with a battalion's eight companies split between a firing line, supports ready to thicken it up or protect its flanks, and a reserve. The tactical manual issued in 1877 was the first to adopt such manoeuvres, although warfare against savage enemies often required a return to close-order formations such as the square.

DRESS 1800–1815

Although this short period raised the prestige of the British Infantry to new heights, it is less important from a dress point of view since the main features, with one important exception, had been fixed in 1796–1797. No Royal warrant was published, but in 1802 clothing regulations were issued to consolidate the changes which had occurred since the 1768 warrant. In December 1811 a general order regulating officers' dress appeared. The period also confirmed a trend that had been emerging for some years: the complete divergence of military fashion from civilian styles.

The important innovation was the 1800 change from the hat to the shako or cap for Guards and Line. The first pattern, nicknamed 'stove-pipe' and fitted with a peak, was of lacquered material, changing to felt in 1806, with a short plume fitted behind a black cockade in front, below which was a large brass plate bearing the Royal cypher within the garter, the crown above, the Royal crest below and trophies at the sides. Special badges were placed within the garter for regiments so entitled and regimental numbers at varying places on the plate. The plume was white over red for battalion companies, green for Light Infantry, and white for Grenadiers, when the latter wore shakos instead of bearskin caps which were reserved largely for parade wear. The metal plate of the bearskin changed to brass, and the bearskin later acquired a peak and white cords with tassels. Fusiliers also had both headdresses. Light Infantry regiments adopted a stringed bugle-horn in metal instead of the large cap plate.

A new, lighter-pattern shako came into use from 1812 with plume and cockade at the left side, a raised front, smaller plate bearing the cypher, number and badges if entitled, and a white worsted plaited cord (green for light companies) suspended across the front with two tassels hanging at the right side. An oilskin cover was issued with this shako. In 1814 all Light Infantry regiments and companies were ordered to wear a stringed bugle-horn and regimental number

From left: battalion sergeant, Line regiment; grenadier, Coldstream Guards; Royal Artillery; battalion private, Line regiment. Engraving after J. A. Atkinson from *A View of the British Army in the year 1803*.

Officer, 7th Royal Fusiliers, 1800. The coat lapels are buttoned back.

instead of the plate; it would appear however that Light Infantry regiments retained the 1806 shako.

Hair continued to be queued, or plaited, for flank companies until 1808; thereafter it was to be cut short.

The coats remained largely as in 1797 but acquired shorter skirts at the rear, with only the forward, outer edges turned back. They also became shorter at the waist. White or buff breeches with knee–length black gaiters continued in use although on service white trousers were worn, at first as overalls, and then instead of breeches. From 1812 grey trousers were more common, with short gaiters worn underneath. Long white gaiters were still retained by the Foot Guards for full dress.

Accoutrements changed little except that a rectangular knapsack replaced the folding pattern. Rolled on top of the knapsack was the greatcoat, first authorised in 1802. It was made of grey kersey, had an extra cape over the shoulders and fell to the calf. From 1811 the collar and cuffs were in the facing colour for most Line sergeants.

Only Light Infantry officers adopted the shako, with green feather and bugle-horn. Field, staff and company officers retained the hat. This had become entirely a bicorne, increasingly worn fore-and-aft, with a crimson and gold cord round the crown, terminating at each point in tufts. It had a feather hackle matching the men's cap plumes and fitted behind a black cockade with a button and gold or silver lace loop. Grenadier and Fusilier officers wore their bearskin caps with gilt plates, or hats with white feathers when their men wore shakos. In the Peninsular War many officers

Plate 11: 1800–1815

31. Chosen Man, The Rifle Corps, 1802. 32. Corporal, 52nd (Oxfordshire Light Infantry) Regiment, 1807. 33. Battalion Officer, 1st Foot Guards, 1808. Plates 11 and 12, of the Napoleonic War, show the alterations effected by the 1802 regulations. Armed with the Baker rifle, the Rifle Corps (Fig. 31), later 95th was the first complete regiment to forgo the red coat, although Rifle battalions of the 60th had adopted green in 1797. This sombre dress had white piping and buttons arranged in three rows down the front of the single-breasted jacket. The designation 'Chosen Man', a superior private, denoted by the white armband, was peculiar to the Rifles. The black accoutrements include the pouch and belt with powder horn and leather case for the forage cap, a waistbelt, with expense pouch on the right front for easier loading, picker and brush suspended by brass chains, and sword-bayonet frog on the left. Normal infantry dress is shown in Fig. 32, the first entire regiment to be designated Light Infantry. Fig. 32's shako bears a bugle-horn instead of the regulation cap plate. Chevrons were first instituted as rank badges in 1803. Fig. 33 wears the hat in its final, bicorne, shape. His lapels are partially buttoned across, showing the tops. White pantaloons were worn with Hessian boots, instead of the regulation breeches and black gaiters, or white for full dress.

31. Chosen Man, Rifle Corps, 1802.

32. Cpl, 52nd L.I., 1807.

33. Offr, 1st Guards, 1808.

34. Dmr, 66th Regt, 1811.

36. Pte, 79th Hldrs, 1815.

35. Sgt, Coldstream Guards, 1815.

Officers of the Rifle Corps (later 95th) and Marines, 1801. Rifles officers had a dress of Light Dragoon style but adopted a cap instead of the helmet soon after 1804. The Marine's uniform, similar to the Line's, has the coat lapels buttoned across, except at the top.

had cut-down hats, sometimes covered in oilskin. From 1812 all officers adopted the new shako ordered for the men. Thus, after 150 years, the hat, in its final form, disappeared as the headdress of the Infantry.

Officers' coats, in scarlet rather than in red like their men's coats, remained as ordered in 1797 with long skirts at the rear, except for Light Infantry who had short skirts like the men's and small buttons all over, making it in effect a jacket. The distribution of epaulettes was unchanged, but they were now edged with 'rich Bullion and Fringe'. Flank companies had 'Scarlet Wings with Bullion and Fringe besides Epaulettes'. From 1811 all officers had bullion and field officers' epaulettes were further distinguished: colonels having a crown and star, lieutenant-colonels a crown, and majors a star.

The official lower garments continued to be white or buff breeches with black-topped boots for field officers and adjutants, gaiters for others; but in practice, boots, usually of the Hessian type, were worn in preference to gaiters. Dark blue pantaloons were permitted in marching order.

From 1811 all officers were to have a jacket, the long coats being kept for Court appearances. On foreign service grey pantaloons or overalls were to be worn; the latter with either short boots or shoes and short gaiters. In 1802 officers had a dark blue double-breasted greatcoat with scarlet falling collar, but from 1811 this was changed to grey, with stand-up collar and cape.

Sergeants were to have scarlet coats in the

Plate 12: 1800–1815

34. Battalion Drummer, 66th (Berkshire) Regiment, 1811. 35. Battalion Sergeant, Coldstream Guards, 1815. 36. Battalion Private, 79th (Cameron Highlanders) Regiment, 1815. Drummers still wore reversed coats and bearskin caps, similar to grenadiers' caps, though, on service, shakos with the universal cap plate were more usual. White and green plumes distinguished the Grenadier and Light companies respectively. Trousers were customary in the field, either white or blue-grey as in Fig. 35, or green for Rifles. Fig. 35 wears the Infantry's shortest-lived headdress, the 1812 shako with plume at the side, shown here in an oilskin foul-weather cover; he also wears the greatcoat introduced in 1803. Battalion company sergeants had carried pikes as a mark of their rank since the abolition of halberts in 1792; in flank companies they carried muskets or fusils. On top of the knapsack, was a mess-tin in a canvas cover. In Highland regiments (Fig. 36) the ostrich feathers now covered most of the bonnet underneath, and the little kilt had replaced the belted plaid. Their jackets were slightly shorter, having eight instead of ten buttons in front. The bonnet peak and grey gaiters were worn on service. The brush and picker were now suspended from the shoulder belt plate. The haversack and water bottle on the left balanced the pouch on the right. Hair of all ranks was cut short from 1808.

Light company officer, 36th Regiment, c 1810. Note the square-cut peak of the cap, only worn by Light Infantry officers. Painting by James Northcote.

sergeants carried fusils and bayonets instead of pikes.

Drummers retained bearskin caps with the King's crest in brass on a black metal plate with appropriate trophies, but often wore their respective companies' shakos on service. The style of their coats followed the men's and the colouring was largely unchanged, except that the white coats of red and black faced regiments were now lined red but faced scarlet — a colour adopted for the facings of all non-Royal regiments. White and buff faced regiments also had red linings, all others white. Guards and Royal regiments' drummers still wore red faced blue with white linings, the former being heavily laced with blue and white, with a white silk fringe round the collar. The coat seams of all other drummers were to be laced but the lace bars on the sleeves were at the colonel's discretion. Those regiments which had red breeches under the 1768 warrant changed to white or buff breeches. Drummers' swords were altered to a 24 inch straight-bladed sword with brass hilts, like sergeants' swords which were of the same pattern as the 1796 officers' pattern.

The feathers on Highland regiments' bonnets increased in number and size, being fastened to a

1802 regulations but otherwise underwent little change. In 1803, chevrons were introduced for all NCOs. Each chevron was to be formed of a double row of regimental lace, edged with a strip in the facing colour and worn, points down, on the upper right arm as follows: sergeant-majors and quartermaster-sergeants four, sergeants three, corporals two. Flank company NCOs had formerly worn their epaulettes on both shoulders, and in due course would wear chevrons on both arms, though the date of authorisation for doing so is uncertain. It was not ordered in the 1803 instruction. In 1813 the senior sergeant of a company was designated 'colour-sergeant', with a special badge worn on the right arm (see Fig. 41, Plate 14). Guards sergeants had gold-laced coats and sashes as follows: 1st Guards crimson with white stripe; Coldstream plain crimson; 3rd crimson, white and blue. All Light Infantry

Battalion officer and men in greatcoats and the 1812 shako. From C. Hamilton Smith's *Costume of the British Army*.

Grenadier, 3rd Guards and Light company private, 5th Regiment, 1815. Note the kidney-shaped mess-tin strapped to the grenadier's knapsack. Introduced in 1814, this type of mess-tin remained in service for some 120 years. From a contemporary French engraving.

wire cage so that they almost concealed the bonnet underneath. Plumes in the same colours as those on the shakos were worn on the left side; the 42nd having been granted a scarlet hackle-feather in 1795, had their flank companies' hackles tipped with this colour. Highland jackets were the same as for English regiments, but shorter, having only eight instead of ten buttons in front. By the end of the eighteenth century the bulky belted plaid had given way to the 'little kilt', formerly worn only in undress, with a length of tartan known as a fly plaid affixed to the left shoulder. On service the fly plaid and sporran were discarded by the men, while officers mainly wore grey or dark blue pantaloons with boots or trousers. On becoming a Light Infantry corps, the 71st lost their Highland dress but wore their diced bonnets blocked out to resemble the shako. Feather bonnets had peaks attached on service, and short grey gaiters were worn over the hose. Officers' and sergeants' sashes went over the left shoulder instead of round the waist.

The formation of Rifles units introduced a new colour into Infantry uniforms, described in the 1802 regulations as 'Dark Green'. A soldier of the Rifle Corps in 1802 is shown in Fig. 31, Plate 11. Pantaloons were ordered instead of breeches, although these were later covered or replaced by

Private, 95th Rifles and officer, 52nd Light Infantry, 1815. From a contemporary French engraving.

trousers on service. The green jackets of the 5th and 6th Battalions of the 60th Foot were faced and piped red respectively and had only a single row of buttons in front; their pantaloons were blue and white respectively.

Rifles officers had a long-skirted, lapelled green coat for dress occasions, but their service uniform drew its inspiration from that of the Light Dragoons, including, at first, their fur-crested helmets. The latter however soon gave way to a felt cap with moveable peak and green cap-lines. Their jackets had three rows of buttons in front, each of twenty-two silver buttons, with black silk twist loopings connecting the buttons, decreasing in length from 7½ in. on either side at the top to 2½ in. at the bottom. No mention is made in the 1802 regulations of a pelisse, but these seem to have been in use by 1804; it was trimmed with brown fur and fastened by black loopings and olivets. A hussar-type crimson sash, green pantaloons and half-boots of Hessian style completed the costume. Their accoutrements consisted of a black leather pouch belt with a whistle and silver ornaments, and narrow black waistbelt with snake clasp and slings supporting a curved sabre. Field officers had a black leather sabretache.

Sergeants were dressed like the men, having a scarlet waist sash with black stripe and white chevrons. They had the riflemen's accoutrements with a 'green ivory' whistle attached, and carried a rifle and sword bayonet.

Instead of drummers, Rifles had buglers (as did

the Light Infantry) whose jackets were as the men's, but with white feathering on the seams, and black and white worsted fringes on the end of the shoulder straps. Their bugles were brass with green cords. Their arms were a sword-bayonet suspended in a frog from a waistbelt like the men's.

DRESS 1816–1828

During the peace that followed Waterloo, uniform became more elaborate, impracticable and, for officers, more costly. In 1822 the first *Officers Dress Regulations* were published. No such regulations were issued for soldiers' clothing, but the design of each item was considered by a Consolidated Board of General Officers prior to the lodging of a sealed pattern. Any changes to dress were promulgated by the issue of *Horse Guards Circular Memoranda*. A soldier's complete kit was divided into 'Clothing', 'Accoutrements' and 'Necessaries'; the latter had to be kept up at the soldier's expense and included such things as his under clothing and his cleaning kit. The scale for all three categories was governed by the periodic issue of Royal warrants.

The short-lived 1812 shako was replaced, from August 1815, by the so-called 'Prussian' model, made of felt, about 7½ in. in height, with a broad top about 9 in. in diameter. It had a polished leather peak, 2½ in. deep, and another, 1¼ in., at the rear; the rear peak was discontinued in 1822 but another inch was added to the shako's height. The men's pattern had bands of yellow or white lace round the base and top, ¾ in. and 2½ in. wide respectively. Their worsted plumes, about 6 in. high (in the same colours as before), fitted into a metal socket placed behind a black leather cockade with regimental button on the upper band. There was a brass crown above a circular dome bearing the regimental number on the front of the shako. The dome was upon a circle of lace and connected to the cockade by brass scales. Light companies and regiments and the Rifle Brigade had a stringed bugle-horn with the regimental number, (or battalion number in the case of the Rifle Brigade) on the cockade. Brass chin scales were attached to lion's head ornaments; they either tied under the chin or were looped up to the cockade. An oilskin cover was issued.

The Coldstream and 3rd Guards adopted similar shakos for their battalion companies, but without the lace bands for the men and with a

Coldstream Guards, 1821. From left: Grenadier officer, marching order; Light and Grenadier company sergeants, Court guard order; adjutant, parade order; Grenadier private (holding horse); battalion officer, Court guard order; battalion sergeant, parade order. Note the slash cuffs of the NCOs, which were not adopted by Guards officers or the Line until 1829. Watercolour by Denis Dighton.

garter star and thistle star respectively. The 1st Guards, having been designated Grenadier Guards in 1815, adopted bearskin caps for the whole regiment, with the usual white plume of grenadiers on the left side.

These bearskin caps continued to be worn for parade purposes by all Grenadier companies and the Fusiliers, but were now higher and broader with a short peak attached to the brass plate (see Fig. 40, Plate 14). From 1824, the wearing of bearskins was confined to the United Kingdom and North America.

The men's coats continued as before, until 1820, when long skirts at the back were ordered for all except Light Infantry. In 1826 a lace loop was placed on each side of the collar and the chest loops were graduated from 5½ in. at the top to 2½ in. at the bottom, giving a more waisted effect.

The old breeches and black gaiters remained as full dress wear until 1823 when they were abolished in favour of blue-grey cloth trousers worn over half-boots, with white linen trousers for ceremonial occasions. The long white gaiters of the Guards disappeared at the same time.

The accoutrements were little changed, except for minor improvements to the pouch, widening the belts to 2¾ in. in 1824 and the approval of a lighter knapsack in 1827.

On the officers' shakos, the lace was gold or

silver, the plume a hackle-feather 12 in. high; Light companies and regiments had green cord cap lines. The lace circle behind the plate and the scales above were discontinued in 1819. The battle honours and special devices which had been mounted on these ornaments were transferred, in 1822, to a new star plate. Light Infantry officers retained the bugle-horn of their men.

In 1816, officers were to wear their coats buttoned across to conceal the lapels, but, from 1822 onwards, coats became a vehicle for extravagant display. The shoulders and chests were padded, the now buttoned-back lapels were cut in a plastron shape and closed with hooks and eyes; lace loops appeared on the 3-in. high Prussian collar and in decreasing lengths on the lapels, on the cuffs and pocket flaps, and on each side of the skirt division behind. At the base of the skirts, where the white turnbacks joined, battalion officers had a regimental ornament, the Grenadiers had a grenade and the Light Infantry had a bugle, embroidered in gold or silver. The epaulettes were embellished with lace, embroidery, beading and bullion fringes, while the flank companies' wings were decorated with chain, beading and bullion plus embroidered grenades or bugles. From 1826 the coats of all, including Light Infantry, had swallow-tail skirts. In undress, officers were to wear either their coats with the lapels buttoned across or a blue single-breasted frock coat. The crimson silk sash was wound twice round the waist, and terminated in bullion fringe ends for battalion companies and in cords and tassels for flank company officers.

Officers' trousers followed the same regulations as the men's, except that white pantaloons and Hessian boots were permitted in Levée dress. From 1823, field officers and adjutants adopted overalls instead of pantaloons and knee boots.

In 1822 a new pattern sword was introduced, with a gilt half-basket hilt, containing the Royal cypher between the outer bars and lined with black leather. The scabbard was black leather with gilt mounts. For company officers this was suspended in a buff leather shoulder belt, 3 in. wide, with frog and belt plate; the belt plate, both for officers and men, was now usually rectangular, instead of oval as formerly. Regimental staff had a black leather waistbelt with slings. A similar belt was introduced, in 1823, for all officers in undress, with a sliding frog for company officers.

Sergeants' and drummers' dress underwent the same changes as the men's but with their distinctions unchanged.

The prevailing mood of over-embellishment also affected the Highland regiments. This was typified by the growing height of their feather bonnets, by ornate dirks, cairngorm shoulder brooches (once only worn by women), and the use of animal skins adorned with gilt or silver bells and tassels to decorate the once-simple leather sporran; as an actual purse this was no larger than it had been, but its hairy covering now spread over a large portion of the front of the kilt. Much of this was doubtless due to George IV's visit to Scotland in 1822, after which Highland costume,

---

Plate 13: 1816–1828

37. Battalion Officer, 80th (Staffordshire Volunteers) Regiment, 1817 (India) 38. Grenadier Private, 59th (2nd Nottinghamshire) Regiment, 1826 (India). 39. Officer, 2nd Battalion, 1st (The Royal) Regiment, c 1825 (India). This plate incorporates the changes after 1815 and shows regiments in India, although there are few concessions to climate other than the light-weight trousers in nankeen, cotton or linen in varying shades of blue. The neat 1812 shako has been replaced by the ornate so-called 'Prussian' pattern of 1815 (Fig. 37), worn with an oilskin cover in the field, as in Fig. 38. The coats of all ranks were generally smartened up, after the looser-fitting garments of the war years, but otherwise were little changed at first. Fig. 38 has his lace in bastion pattern, as adopted by various regiments. Officers' lapels became shaped, to give a plastron effect when buttoned back (not across as in Fig. 33) and the coats were padded to give a waisted look. So costly and impractical did they become that, particularly on foreign service, officers took to wearing undress uniform. The shell jacket of Fig. 39, based on sketches of the 1st Regiment at Madras, did not become regulation for another four years, the normal undress being the blue frock coat. The forage cap and loose trousers are of purely regimental pattern. Soldiers in the field in India usually had their knapsacks carried on regimental transport, hence the man from the 59th, based on an engraving of the Siege of Bhurtpore, is only accoutred with his pouch belt and bayonet belt.

37. Offr, 80th Regt, 1817.

39. Offr, 1st Regt, 1825.

38. Pte, 59th Regt, 1826.

40. Pte, 17th Regt, 1826.

42. Sgt, 60th Rifles, 1828.

41. C/Sgt, 94th Regt, 1828.

in abeyance, except in the Army, since its proscription in 1747, developed from a once workmanlike garb into something approaching fancy dress.

Although the jackets of Highland regiments underwent the same changes as the rest of the Infantry, their jacket skirts continued short as more appropriate over the kilt. From 1822 the jackets of all companies acquired wings.

The kilted regiments were now the 42nd, 78th, 79th, 92nd and 93rd. The 78th wore the so-called 'Mackenzie' tartan, which was merely the Government sett worn by the 42nd with red and white lines added, as was the 92nd's, but with yellow lines added. The 93rd wore the Government tartan but in a slightly lighter shade. Only the 79th had a tartan peculiarly its own, styled the 'Cameron-Erracht'. In 1823 the once-kilted 71st and 72nd were permitted tartan trousers, described generally (though not entirely accurately) in military terminology as trews; the 71st's were in Mackenzie, the 72nd's in the predominantly red 'Prince Charles Edward Stuart', similar to Royal Stuart. The 72nd were also granted feather bonnets, but the 71st, as a Light Infantry corps, continued to wear their diced bonnets blocked to the shape of the current shako.

Highland officers, sergeants, pipers and drummers all retained the basket-hilted broadsword. The hilt was gilt or brass until 1828 when it was changed to steel, except for pipers and drummers. Highland officers' shoulder belts had slings instead of a frog.

Although retaining the sombre colours and main features of the former period even the more functional uniforms of the Rifles did not escape the excesses of this period. On becoming entirely a Rifle regiment the 60th discarded the red coat, worn since 1755 except by its Rifle battalions, and adopted a uniform similar to the Rifle Brigade's but with red facings. The officers' dress of both regiments followed the hussar style except for their shakos, which were of the current pattern but with a falling plume of cock's feathers, the men having horsehair.

Under the 1822 regulations, Rifles officers had the same sword as other regiments. However,

17th Regiment, 1828. Left: drum-major, full dress; white coat, faced red. Right: bugler, Light company, undress; white jacket, green wing tufts. Lithograph after E. Hull

Plate 14: 1816–1828

40. Grenadier Private, 17th (Leicestershire) Regiment, 1826. 41. Battalion Colour-Sergeant, 94th Regiment, 1828. 42. Sergeant, 60th Duke of York's Own Rifle Corps, 1828. This plate represents regiments at home in the extravagant, post-Waterloo period. The grenadier cap (Fig. 40) is larger, and has acquired a peak and chin scales, while the coat has long skirts again, with white turnbacks. The cuffs are still those prescribed by the 1768 warrant. Fig. 41, in walking-out dress, wears the 1815 soldiers' pattern shako with top-heavy plume and chin scales tied above the peak, and wears the white linen trousers worn in summer, instead of the blue-grey cloth, winter variety shown in Fig. 40. Trousers replaced breeches and gaiters in 1823. On the right arm is the special badge instituted for this rank in 1813; flank company colour-sergeants wore a three-bar chevron on the left arm. Fig. 42 shows the 60th's uniform after its entire conversion to a Rifle corps in 1820. It is similar to the Rifle Brigade's (formerly 95th) but with red facings. 60th officers also adopted a jacket and pelisse as worn by officers of the other Rifle regiment since 1800. The falling plume of the shako was of cock's feathers for officers, horsehair for men. A whistle with chain was worn on the pouch belt by both officers and sergeants.

from 1827 they were authorised to have a similar-pattern sword but with steel guard and mounts, a crowned bugle-horn within the outer bars, and a black leather scabbard with steel mounts for full dress, a steel scabbard in undress. The sword knot was black leather.

DRESS 1829–1843

The increasing cost of officers' uniforms, due to regiments adopting embellishments in excess of those ordered by the 1822 regulations, resulted in restraining action being taken by the Horse Guards. An entirely new dress was ordered by *Circular Memorandum* in 1829, its provisions regarding officers being codified in the 1831 Dress Regulations and further elaborated upon in the 1834 regulations.

Except for regimental distinctions and the adoption of bearskin caps by the Grenadier Guards, the dress of the Foot Guards had hitherto approximated to dress of the Line. From 1831, however, it changed markedly when it was decided to accord the Coldstream and 3rd Guards similar status to the Grenadiers by granting them the dress and title of 'Fusiliers'. The Coldstream declined the title but the 3rd Guards were re-designated Scots Fusilier Guards. Both regiments adopted bearskin caps, with a scarlet feather on the right side for the Coldstream but no feather for the latter.

The new upper garment, ordered in 1829, was the coatee. It was double-breasted, without lapels, its two rows of buttons being 3 in. apart at the top, $2\frac{1}{2}$ in. apart at the bottom. It had a Prussian collar and round cuffs in the facing colour (the latter with a buttoned flap, or slash), and swallow-tail skirts with white turnbacks and pocket flaps similar to those on the cuffs. Lace was confined to the collar, cuff slash and pockets. Laced epaulettes with a stripe in the facing colour, metal crescent and bullion fringe were worn on both shoulders by all officers, rank being distinguished by the depth of the fringe and special devices. Trousers were of two types: a very dark almost black shade known as Oxford mixture, to be worn from 15 October to 30 April, and white linen for the rest of the year.

In the Guards the double-breasted coatee was worn by all ranks, with the buttons (gilt for officers, pewter for men) spaced regular, in pairs or threes according to regiment. Each end of an officer's collar was embroidered in gold, with, thereon in silver, a grenade for the Grenadiers, a garter star for the Coldstream and a thistle for the Scots Fusiliers. Similar badges were worn on the epaulettes and as skirt ornaments. The sergeants' coatees resembled the officers', but the rank and file's coatees, though of the same cut, were red with unlaced collars, and the badges, cuff and pocket lace were in white worsted; sergeants and below of the Scots Fusiliers wore the St Andrew's cross instead of a thistle. The rank and file's shoulder straps were plain white, ending in a crescent with a white fringe. From 1833 a $1\frac{1}{2}$ in. scarlet stripe was added to the Oxford mixture

---

Plate 15: 1829–1843

43. Company Officer, 87th Prince of Wales's Own Irish Fusiliers, 1833. 44. Private, 71st (Highland Light Infantry) Regiment, 1837. 45. Bugler, 68th (Durham Light Infantry) Regiment, 1840. The changes of 1829–1830, designed to simplify and economise, resulted in a more elegant, if hardly more practical uniform. The officers' lapelled coat was superseded by the double-breasted swallow-tailed coatee (Fig. 43), with slashed cuffs, fringed epaulettes on both shoulders, or wings for flank companies, Fusiliers, Light Infantry and Highlanders (except for field officers). Fusiliers' and Grenadiers' bearskin caps were simplified. The 1829 shako began with a plume and star plate, later changing to a ball tuft and, for the men, a smaller plate as in Fig. 45. The men's coatee remained single-breasted, although from 1836, the sergeants' coatee was double-breasted like the officers' coatee. The lace loops, varying in length from 1826, changed from coloured to plain white in 1836, except for drummers and buglers (Fig. 45) who kept the coloured lace but lost their reversed coats. Fig. 45 wears the Oxford mixture winter trousers. In 1834 the 71st (Fig. 44) adopted Mackenzie tartan trousers, and their peaked, diced blue bonnets were blocked to the shape of the shako and fitted with bugle badge and black cap lines. The pouch and bayonet were now hung so that neither was visible from the front. The mess-tin was strapped to the knapsack and the greatcoat folded against the back. All belt plates were now rectangular.

43. Offr, 87th Fusiliers, 1833.

44. Pte, 71st L.I., 1837.

45. Bglr, 68th L.I., 1840.

46. Pte, 75th Regt, 1835.

48. Pte, 13th L.I., 1842.

47. Cpl, Grenadier Guards, 1838.

Guards, 1832. Top, from left: Coldstream Guards: Pioneer Corporal Swiffling, Colour-Sergeant Maundrell, Lieutenant and Captain Hon. Thomas Ashburnham, Private Cory. Bottom, from left: Scots Fusilier Guards: Drummer Cann, Colour-Sergeant McDonald, Lieutenant and Captain Hon. Henry Montagu (Adjutant), Private Godfrey. Painting by A. J. Dubois Drahonet.

Plate 16: 1829–1843

46. Grenadier Private, 75th Regiment, 1835 (South Africa). 47. Corporal, 2nd Battalion, Grenadier Guards, 1838 (Canada). 48. Private, 13th (1st Somersetshire Light Infantry) Regiment, 1842 (Afghanistan). This plate illustrates examples of how the regulation uniform was adapted during the early colonial campaigns that were to dominate much of the nineteenth century, especially the increasing use of undress clothing in the field. For bush-fighting in the 6th Kaffir War, the 75th (Fig. 46) wore the blue forage cap with band in the facing colour, fitted with a peak in front and another folding peak behind. The wings and tails of the coatee were removed and leather trousers were adopted; boots were of untanned leather. Waistbelts and pouches of untanned leather were acquired locally. Fig. 47, based on a contemporary watercolour, illustrates a Grenadier Guards NCO, in the Canadian Rebellion, wearing the double-breasted coatee worn by all ranks of the Foot Guards from 1831. Over the red-banded, peaked forage cap is a fur cover, and the winter trousers are protected by blanket material. On his right forearm are two of the good conduct stripes introduced in 1836. Fig. 48 is based on a Cunliffe painting of the 13th in the First Afghan War. The forage cap is green, for a Light Infantry regiment, with band matching the facings and bugle badge. The undress shell jacket is worn with the nankeen trousers issued in India.

Line, 1832. Top, from left: Sergeant Jameson, 2nd Regiment, Captain H. K. Bloomfield, 11th Regiment, Private Sweeney, 88th Regiment. Bottom, from left: Private Rogers, 90th Light Infantry, Captain W. Flood, 51st Light Infantry, Sergeant Smart, 13th Light Infantry. Painting by A. J. Dubois Drahonet.

trousers of the officers, a red welt was added to the men's trousers.

For the Line, the new dress was accompanied by a completely different shako, authorised at the end of 1828. This was bell-topped, $6\frac{1}{4}$ in. high, 11 in. in diameter across the top, with leather bands round the top and bottom and two leather bands at each side in a V. The new shako had a drooping peak 2 in. deep at the centre. Gilt or brass chin scales were suspended from side ornaments; in 1842 these changed to a chin chain for officers and a plain black leather strap for the men. The Hanoverian black cockade was finally dispensed with, and the front bore a large, universal star plate, surmounted by the crown. On this, for officers, was superimposed the star worn on the previous shako; the men's brass plate had the regimental number within a circle, the whole being struck in one piece. Light Infantry companies and regiments had the bugle-horn and number within the centre of the star. In 1839 the men's star plate changed to a circle 3 in. in diameter, surrounded by an oak and laurel wreath with crown above; within the circle was the

Highlanders, 1832. From left: Private Ritchie, 79th, Captain J. E. Alexander, 42nd, Lance-Sergeant, 92nd. Painting by A. J. Dubois Drahonet.

number, surmounted by a bugle or grenade for flank companies. Plumes, in the same colours as before, were 12 in. high for officers but lower for the men. These changed in 1831 to an 8 in. all-white feather or worsted plume, changing again to a white ball-tuft in 1835. Light companies and regiments changed their green feathers for a ball tuft in 1830.

Grenadier companies and Fusiliers continued to wear bearskin caps at the same stations as before. These lost their brass plates and peaks in 1835, as did those of the Guards, but in 1842 all Line Grenadiers adopted shakos, Fusiliers doing so a year or two later.

Unlike the Guards, the men's coatees in the Line remained single-breasted with pewter buttons, either spaced regular or in pairs, and lace loops across the front as before. From 1836 regimental lace was abolished in favour of plain white. The loops were either square-ended, pointed or bastion, according to regiment. Collars were edged all round with lace and had a loop at either end. The shoulder straps were in the facing colour, edged with lace and terminating, for battalion companies, in a white worsted crescent; for flank companies, Fusiliers and Light Infantry shoulder straps terminated in laced red wings with a white worsted tuft around the outer edge. The cuffs were round with a red slash bearing three or four buttons with lace loops. A similar slash appeared on the pockets set in the swallow-tail skirts. At the join of the white turnbacks, was a regimental button, grenade or bugle. From 1836, Good Conduct badges were introduced: a chevron of single lace, point upwards, worn above the right cuff.

In 1833 the Oxford mixture trousers received a $\frac{1}{4}$ in. red welt. The old flap opening which breeches and trousers had had for years changed to a fly front in 1836. At the beginning of the century, trousers had been ankle length, but now they came well down over the heel and were the same width round the bottom as round the knee, ranging from $18\frac{1}{2}$–$19\frac{1}{2}$ in. according to a man's size.

The accoutrements remained much the same, although the pouch and bayonet now hung behind the hips so that neither were visible from the front. An improved knapsack (though still rectangular in shape) was approved in 1829. With it, the greatcoat was to be carried in three ways: in guard mounting order, folded square on the outside; in light marching order (a reduced kit), within it; in heavy marching order (a full kit), rolled on top. In the two former cases the mess-tin was to be strapped on top, in the latter case it was to be flat against the outside, its top in line with the upper edge of the knapsack, secured by its own strap to the pack's back slings. Its cover was white canvas, changed to black in 1838.

Items like haversacks, water bottles, blankets, cooking pots and billhooks were still classified as

46th Regiment, 1837. From left: ensigns with Queen's and Regimental Colours; officer, undress; regimental surgeon. All in summer dress. Watercolour by M. A. Hayes.

camp equipage and only issued when required. The first two items were slung over the shoulders; blankets were folded square against the back of the knapsack; and the other two, when carried by a proportion of the men, were strapped against the blanket.

From 1839 the old flintlock muskets began to be replaced by the percussion musket (see Appendix 4), although the complete change-over took several years. As the flintlocks were withdrawn, so too were the brushes and pickers which had hung from the shoulder belt plates. A proportion of the caps required to fire the new musket were carried in a leather pouch attached by a ring to the right side of the coatee, the balance being in a tin in the pouch.

Line officers had double-breasted coatees of the pattern already described for the Guards, but with two loops and buttons at each end of the collar, instead of the Guards badges and embroidery, and chain-covered wings for flank companies, Fusiliers and Light Infantry. Field officers of the Fusiliers and Light Infantry wore epaulettes. From 1830 gold lace and buttons became universal for all Regular officers, silver henceforth being confined to the Militia. In the same year, the gorget, which had fallen into disuse, was formally abolished.

Sashes and shoulder belts remained as before, but in 1832 field officers were to have a buff leather waistbelt with slings, fastened by a gilt rectangular clasp. At the same date such officers were to have a brass scabbard, and adjutants were to have a steel scabbard though they reverted to leather in the 1834 dress regulations.

In 1836 sergeants received scarlet double-breasted coatees like the officers' coatees, but with collars, cuffs and epaulettes like the men's, except for the sergeant-major and quartermaster-sergeant who had silver lace where the officers had gold. Their crimson waist sashes had three stripes in the facing colour. All flank company NCOs, as well as Fusiliers and Light Infantry, now wore their chevrons on both arms. A colour-sergeant of this category wore his special badge on the right arm and a three-bar chevron on his left arm. In 1830 the pike was abolished for all sergeants, being replaced by a fusil. This necessitated their wearing a pouch belt, while their swords and bayonets occupied a double frog on the other side. The sergeant-major was armed and accoutred like an officer.

Drummers' dress, too, underwent major changes. In 1831 those of non-Royal regiments lost their long-hallowed reversed coats. Henceforth their coats were to be like the men's, but with the customary seam and sleeve lace which, even after the universal adoption of white lace in 1836, was permitted to be of special regimental pattern. All drummers continued to wear wings, but only those of Grenadier companies retained the bearskin caps until these were abolished. Guards drummers, unlike the rest of their regiments, wore single-breasted coatees, heavily embroidered with their own special lace pattern of blue fleur-de-lys on white.

46th Regiment, 1837. From left: Grenadier and Light company sergeants; Colour-sergeant and sergeant, battalion company; sergeant-major. All in summer dress. The Light company of the 46th was permitted a red ball tuft. These sergeants have not yet received the double-breasted coatees, like the sergeant-major's, ordered in 1836. Watercolour by M. A. Hayes.

Bandsmen, who now formed an authorised element of a regiment, had their dress regulated for the first time in this period. They were to be dressed as the rank and file but with double-breasted white coatees and facings of the regimental colour; if a regiment was faced white, its bandsmen's coatees were faced red. Guards bandsmen, or rather musicians, wore single-breasted scarlet coatees laced gold.

The changes affecting the Line's coatees also applied to Highland regiments, except that their coatees kept the short skirts, including the non-kilted 71st and 72nd. All ranks, other than field officers, continued to wear wings.

Rifle regiments adopted the new Line shako, but with all metal fittings bronzed and with black cap lines, which for officers included a plaited festoon falling across the front. In 1836 the falling plume was replaced by a black ball-tuft.

Although the 60th and Rifle Brigade officers both continued to affect hussar dress with looped jacket and pelisse, the men's clothing of both regiments now diverged. The white metal buttons of the previous uniform were exchanged for black in 1833, but while the red-faced short-skirted jackets of the 60th continued much as before, being single-breasted with three rows of buttons, the Rifle Brigade assumed double-breasted coatees with swallow-tail skirts of the same pattern as the Foot Guards rank and file. The collar and cuffs were plain black without lace, the cuff slash green with three buttons, the turnbacks and shoulder straps black, the latter terminating in a black worsted crescent.

The accoutrements of both regiments still consisted of waistbelt and pouch belt, sergeants wearing a bronzed version of the officers' pouch belt badge on their pouch belt, together with a whistle. From 1837 the Brunswick rifle began to replace the well-tried Baker rifle (see Appendix 4).

Although by this period a large proportion of the Infantry was stationed abroad in the colonies, a soldier's uniform served him equally for home and foreign service, whether the latter was in temperate or tropical climate. Occasionally, garments more suited to the country concerned were acquired locally. The 2nd Foot in India during 1828–1830, for example, had white shako covers, white undress jackets for drills, and blue 'dungaree' trousers for fatigues. For dress parades, however, the normal home-service uniform was worn. On campaign the same applied, as can be

Bugler, Colour-sergeant, field officer and officers of the Rifle Brigade, c 1834. The officers' shakos have a festoon and cap lines in addition to the ball tuft worn by the men. Lithograph after S. Eschauzier.

seen in lithographs of the First Afghan War, where the only concession to active service was the wearing of black or white covers for the shakos. In order to avoid wear and tear on dress uniform and to afford a greater measure of comfort, undress uniform was increasingly worn, particularly on foreign service.

Soldiers had had an undress or working uniform for years but its features had been largely a matter of regimental choice. In the 1829 and ensuing orders it was regulated for the first time. The forage cap was blue, with a band in the facing colour, red for Royal regiments. Light Infantry had green caps with similar bands or red if the regiment was faced green. Peaks were fitted to officers' caps, and sometimes to the men's in the East. In 1829 the dimensions of the 2nd Foot's caps were: the crown 12 in. in diameter, the walls $4\frac{1}{2}$ in. high. In 1834 the coloured bands on officers' caps changed to black silk oak-leaf lace, Royal regiments retaining red. The regimental number in gold (or special badge if so entitled) was worn on the front, the Fusiliers and Light Infantry having a grenade or bugle respectively. In the same year the men received a plain dark blue knitted cap, dark green for Light Infantry, with a tourie on top (white or green for the flank companies, black for the remainder), and a brass number with, additionally, the appropriate badge for flank companies. Some regiments retained the facing colour bands for several years (see Fig. 48, Plate 16). The sergeants' caps had peaks, which, after 1836, were also permitted for the rank and file when serving in India, the West Indies or the

Mediterranean. This plain blue or green cap, called a 'Kilmarnock', was to last for nearly thirty years, though worn in numerous ways and issued in different dimensions.

Every officer had a plain dark blue frock coat with regimental buttons. Initially this had a gold cord on each shoulder but this changed in 1834 to lace-edged blue shoulder straps terminating in brass crescents. For regiments in most foreign stations, a scarlet shell jacket, with small gilt buttons, regimental facings and gold shoulder cords, was ordered. This was a very plain garment without any lace or skirts. Types of shell jacket had been worn unofficially by some regiments overseas for several years (see Fig. 39, Plate 13) but no authorised pattern was approved until 1829.

In the following year the soldiers' white fatigue jacket, which had developed from the eighteenth century sleeved waistcoat, was replaced by a red shell jacket. This had pewter buttons and regimental facings, but the cuffs were round, in contrast to the officers' cuffs which were pointed. In some regiments the flank companies' jackets had plain unlaced wings.

At first officers in undress wore their normal shoulder belts, but this was soon changed to a black leather waistbelt with frog for company officers, slings for field officers.

Guards officers' forage caps had a gold-laced peak, with black bands for the Grenadiers and Coldstream and diced band for the Scots Fusiliers. A grenade, garter star and St Andrew's star, respectively, were on the front. They had a braided blue frock coat of regimental pattern but no shell jacket. The bands of the men's caps were red or white or diced for each regiment. Sergeants of the two senior regiments had gold lace bands, Scots Fusilier sergeants having gold piping round the crown. All Guards NCOs and men had peaks to their forage caps in this period. The red shell jacket was not worn, each regiment retaining the old white jackets.

Highland officers' undress generally followed Line officers' undress, though with diced bands round the forage caps, except in the 42nd. Tartan trews were usually worn in undress instead of the kilt. Although scarlet shell jackets were authorised for officers and staff-sergeants, the remainder kept to the white jacket like the Guards.

Rifles' officers had a shell jacket but no frock coat. The men's undress was of the same pattern as the Line's, their forage caps being plain rifle-green, with red touries for the 60th and black for the Rifle Brigade.

DRESS 1844–1854

Although a fresh set of *Officers Dress Regulations* was published in 1846, this period saw only minor changes to clothing. There was, however, a new headdress and an important alteration to the soldier's accoutrements.

The elegant but top-heavy bell-topped shako gave way to a cylindrical shape, somewhat reminiscent of the 1800 pattern. Called the 'Albert'

Plate 17: 1844–1854

49. Private, 60th King's Royal Rifle Corps, 1848 (India). 50. Company Officer, 29th (Worcestershire) Regiment, 1849 (India). 51. Battalion Private, 74th (Highland) Regiment, 1851 (South Africa). Figs. 49 and 50 are typical of the Sikh Wars. Fig. 49, from an eyewitness sketch, wears the 'Albert' shako with linen cover, dress jacket and cotton trousers. Rifles had always had waistbelts, but many other Queen's regiments in India at this time adopted the H.E.I.C. armies' practice of wearing an additional waistbelt to steady the shoulder belts. Officers usually wore a forage cap, with a cover, as in Fig. 50. The latter has a loose-fitting quilted jacket of cotton, or serge, over his shell jacket and cloth trousers. His sword is suspended by a frog from the undress waistbelt and he carries a haversack. In the 8th Kaffir War the 74th abandoned the coatee or shell jacket, worn by other regiments, in favour of the smock-frock, issued to troops as protective clothing on board ship. Its white canvas was dyed to a suitably nondescript shade which together with the tartan trews, provided a rudimentary form of camouflage as well as being a comfortable costume for the bush. Untanned leather boots and accoutrements were obtained locally, and broad leather peaks were fitted to the undress bonnets. Knapsacks were discarded and a soldier's 'necessaries' were carried in a folded blanket slung from the knapsack straps, with the mess-tin secured on top. During this campaign the Minié rifle was issued to a proportion of the troops in the field.

49. Pte, 60th Rifles, 1848.

50. Offr, 29th Regt, 1849.

51. Pte, 74th Hldrs, 1851.

52. Cpl, 7th Fusiliers, 1854.

53. Offr, Scots Fusilier Gds, 1854.

54. C/Sgt, Rifle Brigade, 1854.

shako after the Prince Consort (who was supposed to have had a hand in its design) the new cap was introduced at the end of 1843. Made of black beaver mounted on felt for officers and of stout felt for the men, it was $6\frac{3}{4}$ in. high and 7 in. in diameter across the top, $\frac{1}{4}$ in. wider at the bottom. The top was leather-covered and turned over, and a $\frac{5}{8}$ in. leather band went round the base. Its novelty lay in its having a fixed peak at the rear as well as at the front, the front peak being $2\frac{3}{8}$ in. deep at the centre, the rear peak being $1\frac{1}{4}$ in. deep. At the sides were rose pattern ornaments, gilt for officers, brass for the men. These secured a gilt chin-chain for officers and staff-sergeants, and a black leather chin strap for the men. On the front, officers had a star plate surmounted by the crown, in the centre of which was a garter containing the regimental number with a laurel and palm wreath around. Battle honours were placed on the wreath and the star's rays. The flank companies and Light Infantry regiments had their usual badges embodied on the plate. Fusiliers, who had lost their bearskin caps, had a large domed grenade instead of the star plate, with regimental distinctions on the ball. Rifles had a bronzed bugle with a black corded boss on the upper edge. The shako was surmounted by the white, green or black ball-tufts of the previous pattern, but from 1846 the white ball was reserved for Grenadier companies and Fusiliers, the battalion companies having two-thirds white, one-third red, the former uppermost. The men's shakos had the crowned circular plate from the previous pattern, Fusiliers and Rifles having badges similar to their officers' badges; the bugles on Rifles caps were blackened and the boss was of tooled leather.

The Guards' bearskins remained unchanged but were reduced four inches in height just before the Crimean War, with the view 'to rendering them better adapted than at present for Field Service'.

The coatees of all ranks were the same as before, except that in 1848 the laced pocket slashes were removed from the skirts. This order did not apply to, or was ignored by, the Guards and Highland regiments. In 1845 sergeants' sashes lost their stripes, changing to plain crimson worsted.

In the same year the white summer trousers were abolished in favour of lavender-grey tweed. This did not affect the Guards or regiments serving in hot climates. However the lavender dye proved so variable that in 1850 the colour was changed to a pure indigo blue. It was during this period that the men's footwear, which hitherto had been made to fit either foot, were now constructed on individual lasts.

The long-established practice of supporting the pouch and bayonet by cross-belts was altered in 1850 when a waistbelt with a fixed frog was introduced. The belt fastened with a circular brass clasp of a type still seen today. The pouch still hung from its shoulder belt but part of its weight was borne by the waistbelt by means of a connecting strap linking the pouch to a brass loop in the centre of the waistbelt at the back. This new arrangement was designed to relieve the shoulder and chest from part of the weight of the ammunition. Like all new equipment it took some years to reach the troops, and many of the regiments that embarked for the Crimean War still had the old cross-belts, including the Guards. When the new waistbelts were issued, the shoul-

---

Plate 18: 1844–1854

52. Corporal, 7th Royal Fusiliers, 1854 (Crimea). 53. Ensign, 1st Battalion, Scots Fusilier Guards, 1854 (Crimea). 54. Colour-Sergeant, Rifle Brigade, 1854 (Crimea). The early stages of the Crimean War were fought in the home service full dress, though many regiments soon discarded their 'Albert' shakos in favour of forage caps. The 1850 waistbelt and frog for the bayonet had now been issued to many, though not all regiments; those so equipped had the small pouches for percussion caps attached to the belt in front, as in Fig. 52. The latter also shows how the men's kit was folded up within the blanket, and strapped together with the greatcoat and mess-tin; an ill-advised measure designed to spare the men the weight of their knapsacks. Except for his rolled cloak and haversack, Fig. 53 is dressed for the Battle of the Alma exactly as if parading in London. The Rifle Brigade wore the normal infantry shako but all NCOs and men had the double-breasted coatee of the Guards cut (Fig. 54). Officers still retained their looped jackets and pelisses. Colour-sergeants wore the special badge shown on the right arm, and the black three-bar chevron on the left. All sergeants wore a red sash round the waist and bronzed badge and whistle on the pouch belt.

Types and different orders of dress of the Grenadier, Coldstream and Scots Fusilier Guards, 1846. Except for a reduction in the height of the bearskin cap, this was the uniform worn in the Crimean War. Lithograph after M. A. Hayes.

der-belt plate was dispensed with. The pouch, including its flap, was now about 8 in. long, $4\frac{3}{4}$ in. high and 3 in. deep. Those of the Guards were slightly larger and had a brass regimental plate on the flap.

The small leather pouch containing a proportion of the percussion caps was now carried in one of two ways: in a slit pocket, set behind the lower lace loops on the right front of the coatee, or sewn on the right front of the waistbelt, if issued. In 1854 the Coldstream Guards had a small buff leather pouch sewn on the front of the pouch belt for this purpose. This method was adopted by other regiments in the next period.

In October 1854 a new knapsack was approved. Made of painted linen with leather-bound corners, its dimensions were $12\frac{1}{2} \times 15 \times 3$ in. Only the mess-tin was strapped on top, the greatcoat being folded flat against the outside.

The era of the smoothbore musket was approaching its end and, from 1851, a musket with rifled barrel, the Minié, began to come into service (see Appendix 4). Four of the five infantry divisions in the Crimea had this new weapon, including both battalions of the Rifle Brigade.

After the waistbelt was introduced, Line sergeants lost the swords they had carried for many years. Swords were retained by Guards sergeants, Line staff-sergeants, drummers and bandsmen; pioneers had a short sword or hanger. Officers' swords were still of the 1822 pattern but received an improved blade in 1845.

No alterations were made to the Highland regiments' dress but in 1846 the once-kilted 74th was permitted to wear Highland-pattern coatees, tartan trousers (of the Government sett, with a white line), plaids for the officers, and a cap similar to that worn by the 71st. Despite the shako changing in 1844, the 71st's cap retained the bell-top shape.

Some alterations to the undress uniform occurred in this short period. In 1849 the sergeants' forage caps lost their peaks, to become the same as the rank and file's caps but of better

Grenadier and Light company officers of the 49th Regiment in 'Albert' shakos, 1844. Note how flank company officers' sashes terminated in cords and tassels looped up to the coatee buttons. Two officers in the background are in undress with frock coats. Lithograph after H. Martens.

quality. From 1851 NCOs and men of kilted regiments wore the dark blue 'Glengarry' bonnet as a forage cap, with a diced band (except for the 79th). This order was ignored by the 42nd and did not apply to the non-kilted 71st, 72nd and 74th, all of whom retained Kilmarnocks.

From 1848 all Line officers were to wear the shell jacket in undress, the blue frock being abolished. As an outer garment Line officers were to have a cloak coat in the same colour as the men's greatcoats. However, four years later the blue frock was again allowed, but not on any parade with troops. The new version was double-breasted with cloth-covered buttons.

An entirely new innovation, at the end of this period, made a significant change to the appearance of the infantryman, who for nearly 200 years had been clean-shaven. In July 1854, following the example of the Crimean regiments, all ranks were permitted to grow moustaches. These had to

Grenadier of the 9th Regiment in forage cap, shell jacket and lavender summer trousers, c 1850. He is equipped with pouch belt, the waistbelt introduced in 1850 to carry the bayonet, and a haversack. His greatcoat, folded in the knapsack straps, is behind his right leg. His cuffs are turned back. Watercolour, possibly by Captain Wilkinson, 9th Regiment.

leave two inches between the corner of the mouth and the side whisker. Beards were grown by the Crimean army, but at home 'the chin, underlip and at least 2 inches of the upper part of the throat, was to be shaved.

In the early battles of the Crimean War all regiments fought in their full dress, though usually discarding their shakos in favour of forage caps. However, in the colonial campaigns of this period, in India, South Africa and New Zealand, undress was more commonly worn. It was often supplemented by locally-acquired items such as a white linen cap or shako cover, lightweight or more hard-wearing trousers, and accoutrements of more suitable construction. In the Kaffir Wars in South Africa, many officers hardly bothered with uniform at all, wearing shooting jackets, corduroy trousers and the like. The dress for such

79th Highlanders, 1854. From left: Sergeant-major, undress; private (*back view*), fatigue dress; sergeant, drill order; sergeant, Light company and piper, review order; pioneer and private, marching order. The 1850 pattern waistbelt with fixed frog is clearly visible on the right-hand figure. Painting by D. Cunliffe.

campaigns was entirely a matter for regimental commanding officers, so that two regiments might appear on the battlefield dressed quite differently. At Chillianwallah in 1848, during the Sikh War, an officer of the 29th (whose dress is shown in Fig. 50, Plate 17) noted: 'The 29th were in undress jackets and forage caps . . . the 24th went into action in full dress, with the inconvenient tall shako.' Both caps and shakos had white quilted covers as a protection against the sun. In some regiments a mixture of dress and undress was worn. The 9th Foot, for example, went through the Sikh War in covered Albert shakos and shell jackets, worn above the bright blue cotton trousers often used in India at this time. In the Kaffir campaigns forage caps were sometimes replaced by broad-brimmed hats or the red stocking caps issued to soldiers as nightcaps. It was in the Eighth Kaffir War that the first serious attempt at a practical service uniform was made, by the 74th Highlanders (see Fig. 51, Plate 17).

The conditions under which these colonial campaigns were fought underlined how inconvenient and unsuitable the regulation clothing was for warfare. Furthermore, by the 1850s, the Army's dress had a distinctly old-fashioned appearance when compared to the uniforms of some European armies, particularly the French, who had already experienced similar colonial expeditions. Regimental officers and military reformers had been advocating changes through books and service journals since the 1830s, but it was not until the early 1850s that the authorities at the Horse Guards began to consider the problem seriously. Eventually a decision was reached and in August 1854 the details of an entirely new dress were announced. This, however, was too late for the regiments sent to the Crimea, who landed in the old shakos and coatees and whose subsequent sufferings are too well-known to need repeating here. The first issues of the new uniform were not made until 1855, and will therefore be considered in the next section.

DRESS 1855–1880

From 1768, when the military coat began to be cut away in front, the soldier's upper garment afforded him less and less protection below the waist, as the skirts were progressively sloped off behind. The new dress authorised in 1854 rectified this defect and its issue the following year heralded the era of the tunic, a garment with skirts all round, inspired by the frock coat.

Although there were numerous changes in detail, the tunic continued as the full dress garment up to 1914, and its general construction provided the basis of the soldier's coat in other orders of dress up to 1939.

In view of the number of changes, however, the era will be considered in separate periods, the first terminating immediately before the 1881 reorganisation of the Infantry.

Officers Dress Regulations were published in 1855, 1857, 1864 and 1874 but these, largely, incorporated changes which had occurred a year or two before their issue. Alterations to the men's dress continued to be promulgated by Horse Guards, and later War Office, General Orders, and the lodging of sealed patterns.

Prior to 1854 the question of a new headdress had been considered in parallel with clothing. Some form of helmet had been advocated in the 1830s, either of a Grecian or Roman type or, more practically, something on the lines of the leather pattern adopted by the Prussian Army in 1842. In the event, another shako was decided upon. It was not unlike the 'Albert', being made of the same materials and having the same ball-tufts, as well as front and rear peaks. The chief difference lay in the back being deeper than the front ($7\frac{1}{8}$ in. to $5\frac{1}{4}$ in.), the top being 1 in. less in diameter than the bottom, giving a tilted effect in the current French style. All ranks had a leather chin strap,

Pioneer Manners, Privates Webster and Lemmen, Grenadier Guards, wearing the 1855 pattern double-breasted tunic. They are armed with the Enfield rifle.

Plate 19: 1855–1880

55. Battalion Private, 57th (West Middlesex) Regiment, 1855. 56. Officer, 5th (Northumberland Fusiliers) Regiment, 1868. 57. Colour-Sergeant, 106th (Bombay Light Infantry) Regiment, 1875. This period witnessed the introduction of the tunic and the last three shakos. The 1855 tunic (Fig. 55) was generously cut and double-breasted, becoming single-breasted a year later. As the jacket acquired a smarter cut, the slashed cuffs (Figs. 55 and 56) changed, in 1868, to a pointed style; at first outlined with white lace, and from 1871 with a trefoil (Fig. 57). The 1855 shako gave way, in 1861, to a lower, blue cloth pattern without the back peak; this in turn was replaced in 1869 by the final type (Fig. 57), which was supplanted by the spiked helmet in 1878 (see Plate 25, Fig. 75). Fusiliers wore a falling white plume on the 1855 and 1861 shakos, changing to a racoonskin cap in 1868 (Fig. 56); Fusilier officers later adopted a bearskin cap. All officers received a waistbelt with slings in 1855, the sash being moved to the left shoulder. Rank badges were worn on the collar. The pouch belt and knapsack equipment lasted until c 1871, when the Valise pattern equipment (Fig. 57) was introduced. Oxford mixture trousers were worn in winter; dark blue in summer, until the latter colour became universal. Fig. 57 also shows the marching gaiters and revised colour-sergeants' arm badge, introduced from 1859 and 1869 respectively.

55. Pte, 57th Regt, 1855.

56. Offr, 5th Fusiliers, 1868.

57. C/Sgt, 106th L.I., 1875.

58. Pte, 52nd L.I., 1857.

60. Offr, 13th L.I., 1858.

59. Sgt, 93rd Hldrs, 1857.

Privates of the 72nd Highlanders, 1856, in the double-breasted Highland doublet with diamond-shaped buttons introduced in 1855. The right-hand man's forage cap bears the badge of the Light company. Note the expense pouch on the right front of the waistbelt. The 72nd was the only Highland regiment to wear feather bonnets with trews.

and wore an eight-pointed star, surmounted by a crown, on the front. In the centre of the plate was a garter encircling the number on a black ground. The officers' caps and plates were naturally more elaborate in detail than the men's caps. Colonels and lieutenant-colonels had two rows of gold lace round the top, majors had one row. The details of the new shako were announced in early 1855. The shako was to be worn by all Infantry, except the Guards (who retained their bearskins, though lower than before) and Highlanders. In 1856, Fusiliers and Light Infantry adopted falling plumes in white and green respectively, except for the 5th Fusiliers who had red over white. Fusiliers kept their grenade badge and the Rifles kept a bronzed bugle.

Like the 'Albert', this shako proved heavy and uncomfortable and in 1861 a lower plainer shako, with only a flat peak in front, was authorised. This was made of cork covered in blue cloth, which was stitched giving a quilted effect. The tilted aspect was even more pronounced, the front being 4 in., the back $6\frac{3}{4}$ in. deep. The star plates underwent minor changes, and the ball-tufts, field officers' lace and falling plumes were as before. In 1866 the Fusiliers again acquired their own distinctive headdress, when they received sealskin caps with a grenade in front and, for the 5th Fusiliers, a regimental plume at the side (see Fig. 56, Plate 19); later, the other Fusilier regiments also acquired plumes.

The 1861 shako was replaced by yet another, more ornamental pattern in 1869 which was to prove the last of its line for the Infantry. Of

Plate 20: 1855–1880

58. Private, 52nd (Oxfordshire Light Infantry) Regiment, 1857 (India). 59. Sergeant, 93rd (Sutherland Highlanders) Regiment, 1857 (India). 60. Field Officer, 13th (1st Somerset) (Prince Albert's Light Infantry) Regiment, 1858 (India). Plates 20-22 illustrate how campaign dress began to diverge from the regulation patterns shown in Plate 19. No foreign service kit had yet been devised for the Army as a whole, other than the white clothing worn during the Indian hot weather. This plate exemplifies the diversity of costume adopted during the Indian Mutiny. The 52nd was the first regiment to dye its white clothing for field service. In Fig. 58 the forage cap is concealed beneath a dyed cover and curtain with a puggaree wrapped round; the shirt is worn outside the trousers. Soda water bottles covered in leather or buckram replaced the wooden keg type. The 93rd (Fig. 59) wore their kilts with brown holland tunics, issued for a campaign in China, from which the regiment had been diverted to India. The feather bonnet linings were removed, and moveable quilted peaks were fitted. Greatcoats were carried rolled over the right shoulder. Fig. 60, based on a portrait, wears a jacket or frock, fastened with loops and olivets of a type then popular with many officers, and dyed trousers, tucked into long supple boots. He carries a spyglass in a leather case and has a revolver holster on the regulation sword belt. The men of this regiment had jackets of a lighter grey, and all ranks wore the 1855 shako with cover and the falling Light Infantry plume tied back over the top.

similar dimensions and materials and with the same ball-tuft, this shako was covered in smooth blue cloth and had a gilt or brass chin chain which could be looped up to a hook at the back. Round the top all officers had two $\frac{1}{4}$ in. lines of gold braid, another line of gold braid at the base, another slanting up each side from the rose side ornaments, and yet another up the back. The men had similar lines of braid but in red and black. A new plate was devised, the crown above a garter encircling the number, stencilled or cut-out, the whole surrounded by a laurel wreath. The number of regiments entitled to special badges now far exceeded the Royal regiments and Six Old Corps, totalling thirty-seven in all. The old-established devices usually went within the garter, the newer ones below.

At first the Light Infantry regiments, which now totalled ten, had this latest shako with a green plume, changing to a ball-tuft in 1874. A dark green shako was introduced in the late 1870s.

Rifles wore the new shako, in rifle-green initially, but in 1871 adopted a black lambskin or

All ranks of the Rifle Brigade, c 1858. Some are in tunics, others in shell jackets, but all wear forage caps except the pioneer (*second from right*) who has the 1855 shako. The officers' sword belts (*extreme left and right*) were soon to be worn under the tunic. The colour-sergeant (*second left, front row*) and sergeants can be identified by the whistles and chains on their pouch belts.

Plate 21: 1855–1880

61. Private, 101st Royal Bengal Fusiliers, 1863 (India). 62. Private, 68th (Durham Light Infantry) Regiment 1863 (New Zealand). 63. Officer, 4th (The King's Own) Regiment, 1868 (Abyssinia). Although various shades of khaki were widely used in the Mutiny, it was discontinued thereafter. For the Umbeyla campaign on the North-West Frontier troops wore the loose-fitting serge frock which had replaced the shell jacket in undress, with serge trousers and forage cap or tropical helmet (Fig. 61); the knapsack would have been transported in the field. Both the 101st and 103rd had red-banded forage caps, a legacy of their former service as H.E.I.C. European regiments. A similar campaign dress prevailed in the Maori Wars of the 1860s but with blue frocks as being more suitable for forest fighting than the red (Fig. 62). Fig. 62 wears Oxford mixture trousers and marching gaiters. In addition to his belts and haversack, he carries a rolled greatcoat tied across the chest by its sleeves, with mess-tin attached.For the Abyssinian expedition, the troops, all from India, wore their white summer clothing dyed khaki, with gaiters and the first pattern tropical helmet with air-pipe. The jacket of Fig. 63 was the white version of the blue patrol jacket which formed part of an officer's undress uniform throughout the later nineteenth century. He carries a haversack, a revolver on his sword belt and a binocular case.

61. Pte, 101st Fusiliers, 1863.

62. Pte, 68th L.I., 1863.

63. Offr, 4th Regt, 1868.

64. Offr, Rifle Brigade, 1874.

66. Sgt, 72nd Hldrs, 1879.

65. Pte, 90th L.I.(M.I.), 1879.

76th Regiment, 1860. From left: bandsman, night sentry, pioneer, officer — undress, ensign, major and lieutenant-colonel (both mounted), quartermaster, captain, surgeon, officer — mess dress, drum-major, Colour-sergeant, sergeant — undress, sergeant-major, drummer. The 1855 shako and the first single-breasted tunic are worn. Lithograph after W. Sharpe.

sealskin busby of similar shape but with no peak. This had a corded boss, bronzed ornaments, a feathered plume for officers and short, upright, horsehair plume for the men.

A year after the introduction of this shako, Prussia defeated France and the latter's military fashions, which had influenced the last three British shakos, lost favour. In 1878, nearly forty years after it was first mooted as an alternative headdress, a helmet was approved for the Infantry. Made of cork and covered with blue cloth (dark green for Light Infantry), this helmet had a spike as did the Prussian model which had inspired its adoption, but its silhouette owed more to the white foreign-service helmet adopted by the British a few years before. The front and rear peaks were rounded for the men, but the officers' pattern was pointed in front and bound with gilt metal, while the rear was deeper, more square-cut and had a gilt bar running from the base up to the

Plate 22: 1855–1880

64. Officer, 2nd Battalion, Rifle Brigade, 1874 (West Africa). 65. Private, 90th (Perthshire Volunteers) Light Infantry, 1879 attached to Mounted Infantry (South Africa). 66. Sergeant, 72nd (Duke of Albany's Own Highlanders) Regiment, 1879 (Afghanistan). Fig. 64 shows the new pattern tropical helmet and the specially designed, light tweed campaign dress with canvas gaiters issued for the Ashanti War. This officer wears one of the new Sam Browne belts, suspending an Elcho bayonet with swelling spear-point blade, which replaced the sword for this campaign. Despite this kit's popularity, it was not developed afterwards. In the Zulu War, the serge frock, issued as undress for the tunic in Plate 19, Fig. 57, was used. In South Africa no puggaree was worn on the helmet which was dyed. Mounted Infantry were increasingly used in that theatre, the men wearing normal infantry dress, but with corduroy breeches, and gaiters (Fig. 65); ammunition was carried in a bandolier. In the Second Afghan War khaki covers were fitted over the white helmets. In winter a khaki Norfolk jacket was worn over the red serge; for summer the white clothing was again dyed. The 72nd, a non-kilted regiment, wore these jackets, with cartridge loops sewn across the front, with their tartan trews (Fig. 66). Despite the introduction of the Valise equipment, several regiments in India at this time still had the old pouch belt accoutrements.

Foot Guards, 1866. From left: Colour-sergeant, Grenadiers; staff sergeant, Coldstream; sergeant-major, Scots Fusiliers; private, Coldstream (greatcoat); pioneer, Grenadiers. With only minor modifications the clothing is the same as that worn today in full dress. Lithograph after J. Ferguson.

Line regiments, 1864. From left: private, 101st Fusiliers (greatcoat); drum-major; pioneer, Fusilier regiment; private, 26th Regiment (undress); Colour-sergeant, Light Infantry regiment; sergeant-major. The third and fifth figures wear the 1855 shako. Watercolour by J. Ferguson.

spike. All ranks had chin chains with rose side-ornaments, a hook at the back behind the spike and a crowned star plate bearing the number and special devices where entitled. Fusiliers retained their special caps, now made of racoonskin. Rifles, however, had to adopt the helmet, covered in rifle-green cloth (in effect black) with bronze fittings, a black Maltese cross and crown for the 60th, a white metal star with badge superimposed for the Rifle Brigade.

Turning from headdress to clothing, the first tunic, unlike the coatee, was of the same basic cut for all ranks. The collar was lower than before, the cuffs were round with a three-buttoned slash, and all, together with the shoulder straps, were in the facing colour, piped white. All lace was abolished, except on the slashes, and the buttons changed from pewter to brass. The first pattern was double-breasted with two rows of nine buttons in front. The skirts were full, and quite long, with slashed flaps at the back. In 1856 this was changed to single-breasted and the skirts made shorter.

The officers' double-breasted tunic had lapels in the facing colour which could be folded back at the top. Epaulettes were abolished, being replaced by a crimson silk twisted cord, on the left shoulder only, to retain the sash which was moved from the waist. Badges of rank appeared on the collar: a crown and star for colonels and captains, a crown for lieutenant-colonels and lieutenants, a star for majors and ensigns. Field officers had gold lace around the top and bottom of the collar, the cuff and skirt slashes, and two rings round the cuff. Company officers had lace round the top of the collar only and one ring on the cuff.

In 1868 the slash flaps on cuffs and skirts were abolished, and the soldiers' cuffs became pointed with the upper edge outlined in white tracing braid and a chevron of white lace just below it. This did not apply to the Foot Guards, whose tunics retained the slash flaps, as they have done to the present day.

From 1871 the white chevron was removed from the cuffs and the tracing braid was looped into a trefoil. The regimental devices on buttons were altered to a universal pattern of the Royal arms. In the following year the colour of the rank and file's tunics was changed to scarlet, making them similar in colour to those worn by the officers and sergeants. The shoulder straps were no longer in the facing colour, and had brass numerals instead of embroidered white ones. In 1874 brass regimental collar badges were introduced; a novelty which compensated for the loss of the regimental buttons three years before.

When the men's cuffs changed in 1868, so did the officers' cuffs, their points being outlined in gold lace with tracing braid forming an Austrian knot, the quantity and arrangement of such

74th Highlanders, 1864. From left: officers, full dress and drill order; private, full dress and corporal, drill order. A similar dress was worn by the 71st and 91st.

74th Highlanders, 1864. From left: pipe-major; band sergeant; private (greatcoat).

decoration varying according to rank. Badges of rank remained on the collar until 1880 when they were placed on twisted gold shoulder cords, as follows, colonels a crown and two stars, lieutenant-colonels a crown and a star, majors a crown, captains and lieutenants two and one stars respectively: ensigns no badge. The latter rank changed to 'sub-lieutenant' in 1871 and to 'second-lieutenant' in 1877.

Officers' sashes were now worn over the left shoulder, sergeants' sashes over the right. From 1868 sergeants' chevrons changed from white to gold lace.

Oxford-mixture trousers continued to be worn in winter but the summer pattern was dark blue with a scarlet welt. White linen trousers were still authorised for summer wear in hot climates, in the 1855 regulations, but only for India in the regulations for 1857. Thereafter white linen trousers ceased to be worn with the scarlet tunic. From 1872 field officers and adjutants were to wear pantaloons and knee boots. In 1859 black leather marching gaiters were introduced. Initially, they were to be worn at the discretion of commanding officers but by the 1870s they were

Infantry officers, c 1875, wearing the last, 1869, pattern shako. The group includes two Fusiliers (*second and sixth from left, standing*); three Highlanders, 78th (*left*), 91st (*centre front*), 93rd (*right*); and one Rifle Brigade (*centre back*). The officers in pillbox caps and pouch belts are all cavalry.

always worn in marching order.

The pouch belt and knapsack accoutrements continued in use, with minor improvements introduced periodically. From 1855 a black leather 'expense' pouch containing twenty rounds of ammunition was worn on the right front of the waistbelt. It was of similar construction to the large rearward pouch, but in 1857 was changed for a pouch of buff leather and different shape. In the same year the haversack which, hitherto, had only been issued occasionally as an item of camp equipage, became a permanent part of the soldier's accoutrements and was to be slung over his right shoulder.

From 1871 entirely new accoutrements came into service. The knapsack was replaced by a black canvas valise which rested on the pelvic area, supported by buff leather braces which crossed between the shoulder blades, passing over the shoulders to rings just in front of each armpit. From these rings, two straps on each side returned to the top and bottom of the valise, while a third strap was fastened to the buff waistbelt in front either side of the clasp. The ammunition was carried in two 20-round leather pouches, either black or buff, fitted on the waistbelt in front. A black leather expense pouch, or ball bag, holding a further thirty rounds hung below the right-hand pouch, occasionally from the back of the waistbelt when the valise was not carried. On top of the valise, and below the greatcoat (which was folded flat and secured to the braces across the shoulders), was the black covered mess-tin. The accoutrements were completed by the bayonet frog at the left hip, underneath the haversack, and a one-quart barrel-shaped wooden water-bottle with a buff strap over the left shoulder. This Valise pattern equipment had its defects

Corporal, 23rd Royal Welch Fusiliers, in 1868 pattern tunic and Valise equipment, c 1872. The pouches shown are of the expense type worn with the previous accoutrements and gave way to a more box-like pattern.

as a load carrier but it represented a major change from the old system, which had lasted in its essentials for so many years, and its design enabled unwanted components to be discarded without disturbing the balance of the remainder. It took time for the new equipment to reach all regiments, particularly those in India, and there is evidence of the old pouch belts being in use as late as 1882.

With the new uniform, officers gave up their shoulder belts with the decorative plates and, henceforth, all officers suspended their swords from waistbelts, fitted with slings and fastened with a circular clasp, which was similar to, but more ornate than the men's. From the Crimean War onwards many officers equipped themselves with a revolver, for which a holster on the waistbelt was necessary. The loss of an arm by an Indian Cavalry officer, Sam Browne, led him to devise a brown leather belt with shoulder brace and frog to carry his sword and revolver. The advantage of such a belt for service, compared with the sling waistbelt, soon made itself apparent to many officers, and during the Second Afghan War it gained wide though unofficial popularity.

The Minié rifle was supplanted by an improved weapon, the Enfield. But the age of the muzzle-loader was drawing to a close and the first breech-loader, the Snider, was introduced in 1866. This, however, was a stop-gap weapon and in 1874 the first of the modern rifles, the Martini-Henry, came into service (see Appendix 4). Since the Martini-Henry was six inches shorter than its predecessors, which had 17 in. bayonets, the lack of reach of the new rifle was compensated by its having a 22 in. bayonet; the men having the familiar socket type, of triangular section, sergeants having a sword bayonet with curved blade.

At first drummers' tunics of the double-breasted type were white, like the bandsmen's, but from 1856 they reverted to red with special regimental lace. In 1866 the special regimental lace was abolished in favour of a universal Line drummers' lace. This was white with red crowns. Drummers of the Foot Guards kept to their fleur-de-lys pattern — which they still wear today. A new type of side drum, based on a French model and much shallower than before, was introduced in 1858 but by the end of this period the old type was reintroduced.

Bandsmen's tunics continued to be white until 1873 when they were changed to scarlet, in conformity with the rank and file. The new tunic had wings, white piping on back and sleeve seams, and a badge of a lyre and crossed trumpets on the upper right arm.

In 1855 the Highland regiments adopted a doublet. Above the waist it followed the pattern of the tunic of the brief double-breasted period, with diamond-shaped buttons. The lower part had Inverness skirts, four separate double flaps, each with three buttons and braid loops. For some years the cuff was the same as the cuff of a tunic but in 1868 a gauntlet cuff with buttons and loops was introduced. All Highland regiments wore this doublet, but quite apart from normal regimental distinctions there were other variations in their clothing: the 42nd, 78th, 79th, 92nd and 93rd were all kilted, with feather bonnets, the 71st, 74th and, from 1864 the 91st, all had trews and diced shakos. The 72nd had trews and feather bonnet. The 91st received the Government tartan with a red line. Highland officers' accoutrements

Pioneer lance-corporal, 93rd Highlanders, in 1868 pattern doublet with gauntlet cuffs and Valise equipment, c 1875. Note the collar badges introduced in 1874; for this regiment an Imperial crown in brass.

Captain, 107th Regiment, c 1880, wearing the 1878 pattern helmet and accoutred for a levée, when a crimson and gold sash and gold-laced waistbelt replaced the usual plain crimson and white articles, and a double gold stripe overlaid the red trouser welts.

Captain Ferguson, Rifle Brigade, c 1875, wearing the first Rifles busby replaced by the helmet in 1878. Note the sword slings are suspended from a belt worn under the tunic.

differed from the rest of the Line's, in that, although they had a gold-laced waistbelt to carry the dirk, the broadsword was suspended from slings attached to the old type of shoulder belt, complete with regimental plate.

In the previous period the pipers of some regiments had worn doublets, and these were now green, usually worn with the Glengarry, though the 42nd's wore the bonnet. The pipers of non-kilted regiments were permitted the kilt.

Rifle regiments adopted the new clothing in 1855, in their distinctive colouring. Their officers' clothing (except for the headdress which has already been discussed) continued to follow the Light Cavalry pattern, which had assumed a skirted tunic with loops and olivets down the front in place of the jacket and pelisse. The sword belt was now worn under the tunic, only the pouch belt showing over it. When the Valise equipment was introduced, the Rifles pattern was entirely of black leather, with the traditional snake clasp for the waistbelt.

The undress uniform also underwent important changes in this period. For a while the men's cap continued to be the Kilmarnock, but from 1868 onwards it began to be replaced by the dark blue Glengarry. This did not apply to the Guards who kept their round forage caps with coloured bands. The Guards also had a dark blue field-service cap, first introduced in 1854, which was flat and had adjustable flaps at the sides. The officers' forage cap remained much as before but with a broad flat peak and, from 1857, a button and trimming on top. In 1864 and 1874 a number of regiments with Scottish, rather than Highland, connections, such as the 21st and 25th, were permitted diced bands. Diced bands were also to be worn on the Glengarry, which was authorised as an additional undress cap for all Line officers 'on active service and peace manoeuvres'.

In 1856 soldiers in hot climates were to have a skirted undress garment known as a frock instead of the shell jacket. From 1870 the frock, made of serge rather than of the cloth of the tunic and with only five buttons in front, replaced the shell jacket for home as well as foreign service. Originally a loose-fitting garment with a flapped breast pocket, it gradually was smartened up with patches in the facing colour on the collar and, in some regiments, on the cuffs as well. Guards and Highlanders, however, still adhered to their white shells for drill purposes, though receiving frocks for manoeuvres and foreign service.

Officers at home retained the shell jacket and double-breasted blue frock coat (now with brass buttons) until 1867 when both were abolished. They were replaced by a dark blue patrol jacket, skirted and fastened with loops and olivets. Officers in India had adopted a scarlet frock roughly similar to the men's in the 1860s, and in 1872 all officers were permitted a scarlet patrol jacket in addition to the blue one. After some rather confusing regulations as to which was to be worn when, the scarlet patrol was abolished, for home service, two years later, though it continued in use in India and other foreign stations. Like the men's frocks it fastened with five buttons, and had a collar of the facing colour.

It was in these undress garments that many of the campaigns of this period were fought, although the Maori Wars of the 1860s saw a blue frock being worn (see Fig. 62, Plate 21). In India, regiments had an all-white clothing for the hot weather season. Since much of the fighting in the Indian Mutiny occurred at this time of year, some regiments rendered this clothing more suitable for action by dyeing it a more serviceable colour — a practice first instituted in India in 1849, when an inconspicuous uniform had been required for the newly formed Corps of Guides on the North-West Frontier. The Mutiny saw the beginning of khaki in the British Army, and clothing in this colour, usually of a greyish hue rather than the greenish-brown now associated with it, was widely adopted during its operations. Nevertheless, many re-inforcing battalions sent out to the Mutiny from

2nd Battalion 8th Regiment, 1879, in the mixed dress of khaki jackets and home service trousers worn during the winter in the Second Afghan War. The battalion is formed in quarter column of double companies, with the officers and Colours in front, the band and drums on the right flank. All the officers have Sam Browne belts.

England, had to fight in their red or rifle-green until they could acquire more suitable clothing. Despite its popularity, khaki was abandoned after the Mutiny and the traditional colours were resumed, even on active service, such as on the North-West Frontier and the Third China War. Not until 1868 did khaki re-appear, when British regiments sent from India to Abyssinia again dyed their white clothing for the campaign (see Fig. 63, Plate 21). The same was done in the Second Afghan War, by the end of which permanently dyed khaki clothing was being issued to troops in the field. Another important dress feature of this campaign was the adoption of puttees, bandage-like strips of cloth to protect the lower trouser legs. These afforded better support for the ankles than the leather gaiters worn at home and elsewhere.

During the Mutiny many officers acquired helmets made of cork, felt, or canvas on wicker frames, in preference to the covered forage cap or shako. From 1860 a cork helmet with an air pipe began to be issued to all regiments in India.

Outside India the undress uniform served for campaigning, although the Ashanti War of 1874 saw the issue of a special clothing made of light Elcho-grey tweed, and a canvas-covered cork helmet (Fig. 64, Plate 22). This clothing was subsequently withdrawn, but in 1877 a white helmet of similar design was authorised for wear on all foreign service. It was to have a puggaree, an additional length of material, tied round it in India, Ceylon, Hong Kong and the Straits Settlements, but not in the Mediterranean, West Indies, Bermuda, the Cape, St Helena, Mauritius, West Africa and Canada. For parade and similar occasions it had a gilt spike and plate, both being removed in undress or on service, when the spike was replaced by a zinc dome covered with white cloth.

DRESS 1881–1901

The abolition of numbered regiments, and the consequent amalgamation by pairs of old regiments to form new ones under territorial designations, necessitated a complete change of badges, shoulder titles and the like. The opportunity was also taken to rationalise the ancient multi-coloured regimental facings on to a national basis: white for English and Welsh regiments, yellow for Scottish regiments, and green for Irish regiments; Royal regiments retained blue. Both the loss of the numbers and the facings caused enormous resentment in the Line Infantry, and from 1890 some regiments managed to regain their old facings. The first *Officers Dress Regulations* to take note of the reorganisation were those of 1883, followed by the 1891, 1894 and finally the fully illustrated 1900 publications.

The Foot Guards were unaffected by the 1881 reorganisation. Since their dress continued as in the previous period it will not be considered further, except to note at the end of this period, the formation of the Irish Guards, whose uniforms were distinguished by buttons in fours and a blue-green plume on the right of the bearskin cap.

The Line's blue helmet remained unchanged, though with different star-plate centres. In the late 1880s an experimental white helmet, with a more flaring brim in the men's version, was issued to certain regiments at home but was withdrawn towards the end of the period, and all, again, resumed the blue type. Fusiliers, now joined by the 20th and 27th as Lancashire and Royal Inniskilling Fusiliers, retained their special caps with distinguishing plumes (see Appendix 2).

The two Rifle regiments had never been happy with the substitution of the helmet for the busby and in 1890 the latter was restored but in a different style. This headdress was also ordered for one of two new Rifle regiments, the Royal Irish Rifles, formed from the 83rd and 86th.

The tunic remained of the same general cut, but with the Royal facings or the new national colours on the collar and cuffs. The cuffs now became round, with no extra embellishment. The shoulder straps were plain scarlet with the abbreviated title or initials of the regiment embroidered in white. Officers kept their pointed cuffs, with the lace and braid according to rank.

The regiments whose dress was most affected by the 1881 reorganisation were four ancient corps of Lowland Scottish origins: Royal Scots, Royal Scots Fusiliers, King's Own Scottish Borderers and Cameronians. As the 1st, 21st, 25th and 26th Foot they had always, appropriately, been dressed as the English, Welsh and Irish infantry, in the same way broadly speaking as had the Scots Guards. Henceforth the Lowland regiments' clothing was to consist primarily of garments worn hitherto only by the vastly junior Highland regiments — the doublet and tartan trews — a change for which there was no historical or geographical justification or indeed much desire on the regiments' part. The Scots Fusiliers in particular resisted most strongly, but their efforts, unlike those of the Scots Guards, were unsuccessful. Somewhat incongruously this Highland costume was to be topped by the helmet for the Royal Scots and King's Own Scottish Borderers; the Fusiliers retaining their fur caps. As for the 26th, their amalgamation with the 90th Light Infantry turned them into the other new Rifle regiment — Cameronians (Scottish Rifles) with rifle-green doublet, tartan trews and helmet. In 1890 the helmet was changed to a green shako, similar to that worn by the Highland Light Infantry (71st and 74th).

Two other regiments underwent a complete dress change. The 73rd and 75th gave up the English uniform, and reverted to the Highland dress they had worn prior to 1809, on becoming the 2nd and 1st Battalions of the Black Watch and Gordon Highlanders. The non-kilted 72nd and 91st had to adopt the uniform of their junior partners, the 78th and 93rd, when they formed the Seaforth Highlanders and Argyll and Sutherland Highlanders respectively. All the tartans assumed by Lowland and Highland regiments can be found in Appendix 2. The 71st and 74th, having shared a similar uniform, suffered less change on becoming the Highland Light Infantry.

Although green facings were prescribed for Irish regiments, this, in fact, only applied to the Connaught Rangers (88th and 94th) as all other Irish regiments had the Royal title and thus wore blue. The new Irish Rifles, though Royal, wore green. The old Rifle regiments, K.R.R.C. and Rifle Brigade, were unaffected by the 1881 changes.

Field officer, 2nd Battalion Connaught Rangers, 1881. Note the plain black leather sabretache worn by mounted infantry officers.

The period saw two major changes in accoutrements. In 1882 a revised Valise equipment was approved. The twin pouches, of buff leather, were slightly larger and more pliable and the ball bag was dispensed with. The greatcoat went into the valise, which became more rectangular and was

---

Plate 23: 1881–1901

67. Field Officer, 60th King's Royal Rifle Corps, 1881 (South Africa). 68. Private, 2nd Battalion, Highland Light Infantry, 1882 (Egypt). 69. Private, 1st Battalion, Princess Charlotte of Wales's (Berkshire Regiment), 1885 (Sudan). The 1881 Infantry reorganisation rationalised facing colours and changed insignia. In the Transvaal and Egyptian Wars the practice of home service clothing for Africa continued except for the foreign service helmet (Figs. 67 and 68). The officer in Fig. 67 wears a serge frock similar to his men's, which, owing to difficulties in obtaining a fast rifle-green dye, was now virtually black. Although Mackenzie tartan was ordered for the Highland Light Infantry, its 2nd Battalion (Fig. 68) fought in Egypt in its old 74th trews. Some contemporary drawings of the campaign show the battalion in the regulation Scottish serge frock (as illustrated overleaf) but in a photograph taken after the war the men are in English frocks, probably issued as a temporary replacement. Unlike in South Africa, a puggaree was worn round the helmet in Egypt and the Sudan. Permanently dyed khaki cotton drill had appeared in the final stages of the Afghan War and the khaki worn by the contingent from India in Egypt gained favourable notice from the authorities. Thus the troops sent out to the Sudan in 1884 were dressed in grey serge frocks and trousers (Fig. 69) instead of the home service clothing. Fig. 69 and Fig. 68 wear a reduced equipment of the first Valise pattern.

67. Offr, 60th Rifles, 1881.

68. Pte, 2nd H.L.I., 1882.

69. Pte, 1st Berkshire Regt, 1885.

70. C/Sgt, Rifle Brigade, 1898.

72. Pte, 2nd Gordon Hldrs, 1901.

71. Offr, 2nd R. Irish Rifles, 1900.

worn higher than before, being centred on the small of the back, the upper-edge level with the armpits. The mess-tin went on top and the water bottle received a flexible bar to clip over the waistbelt, thus dispensing with its cross-strap.

When full the pouches of this equipment hung badly on the waistbelt, and the valise could not be speedily removed without upsetting the balance of the accoutrements. In 1888 the equipment was supplanted by the Slade-Wallace pattern which attempted to rectify these defects. This was also lighter, had fewer straps and was better balanced. Two shoulder braces were attached to the waistbelt in front and at the back, the rear ends supporting the rolled greatcoat. The two latter items balanced the twin pouches in front. The first pattern pouches carried a total of seventy rounds but, with the introduction of the magazine rifle, this increased to ninety rounds, then to a hundred. On the outer edges of each pouch were two loops, so that four rounds were ready for immediate use. From 1894 the pouches were reversed so as to open outwards. The valise, which was smaller than the 1882 pattern, rode high on the shoulders, its straps passing through metal **D**s on the braces just behind the shoulders and fastening to the brace buckles. This valise could be removed in three or four seconds, compared with the ten minutes it took for the 1882 pattern. The haversack was as before, but a lighter circular water bottle with felt covering and separate carrying strap was introduced.

In addition to the white sling waistbelt, the 1891 dress regulations permitted officers to wear 'a brown belt' on active service and certain foreign stations. In 1899 the Sam Browne belt, with double braces, received official sanction for all active service and peace manoeuvres world-wide.

In 1888 the first of the bolt-action magazine rifles, the Lee-Metford of .303 in. calibre, was introduced to replace the single-shot Martini-Henry. This in turn gave way to the improved Lee-Enfield in 1895 (see Appendix 4). Officers' swords remained basically of the 1822 pattern until 1895, when a straight sword with three-quarter basket hilt was authorised. The straight blade had come in three years earlier. Guards and Rifles officers adopted the straight blade but kept the old three-bar hilt in steel. In keeping with the assumption of Highland clothing, Lowland regiments received a broadsword. However, this was only carried in full dress, and on all other occasions the basket hilt was replaced by a simple cross-bar guard. This applied to all Scottish regiments except the Cameronians who had a Rifles sword.

From 1895 drummers, buglers and bandsmen carried a short sword with $13\frac{1}{8}$ in. blade and cruciform hilt; iron for Rifles, brass for the remainder. The first version of this sword had been introduced in 1856 with a 19 in. blade and a more ornate but basically similar hilt. Pipers, drummers and bandsmen of all Scottish regiments, except the Cameronians, were armed with dirks, which had been carried by musicians of Highland regiments since 1871.

In undress, the Glengarry began to be replaced in the mid 1890s by the dark-blue field service

Plate 24: 1881–1901

70. Colour-Sergeant, 2nd Battalion, Rifle Brigade, 1898 (Sudan). 71. Officer, 2nd Battalion, Royal Irish Rifles, 1900 (South Africa). 72. Private, 2nd Battalion, Gordon Highlanders, 1901 (South Africa). Khaki drill was worn for all purposes other than ceremonial in India from 1885, but similar dress for other foreign service was not authorised until 1896. The latter was worn for the first time on active service by troops from home in the Sudan campaign of 1898 (Fig. 70). The trousers were tucked into puttees, first worn in the Afghan War. The white helmet had a khaki cover and curtain with regimental flash on the side. Fig. 70. has the Slade-Wallace equipment, in black leather for Rifles. His haversack is worn across the shoulders, its sling looped under the pouches. The same dress was worn in the opening stages of the South African War but the cotton drill material was soon replaced by warmer and harder wearing serge, as in Fig. 71. This officer, wearing breeches instead of trousers, has Sam Browne equipment with double braces, and carries a rifle, the better to conceal his rank from Boer marksmen. This need for concealment led to the adoption of khaki kilt aprons by Highlanders. Fig. 72 has acquired a slouch hat in preference to the helmet and wears the buff leather Slade-Wallace equipment, without the valise but with greatcoat rolled at the back of the belt.

Privates, Coldstream Guards, in marching order with Slade-Wallace equipment, c 1890. The haversack when not full was rolled up. They are armed with the first magazine rifle, the Lee-Metford.

Privates, King's Royal Rifle Corps in marching order with Slade-Wallace equipment, serge frocks and the helmet worn before the re-introduction of the busby in 1890. The leather pouch above the brace buckle on the right-hand figure contained a spare magazine for the rifle. Their equipment is all black, including the haversack.

cap. This cap had a turned-up peak, and sides which could be let down or fastened up, and two small buttons in front. The officers' forage cap remained similar to the 1880 pattern and the 1894 dress regulations authorised the replacement of the Glengarry by the field service cap. However, from 1885 another type of folding cap, of a pattern introduced in 1877 for staff officers and not unlike the Guards 1854 field cap, had been quite widely worn by Infantry officers, though not so ordered in dress regulations. By the mid 1890s the field service cap was universal.

The soldiers' serge frock instituted in 1882 was virtually indistinguishable from the cloth tunic, except for the material, rounded skirts in front and no piping. A short-lived all-scarlet version, with facing colour on the shoulder straps only, appeared in 1891–1892, but was discontinued thereafter, being replaced by the 1882 pattern again. Regiments in India had similar frocks but with only five buttons in front and plain scarlet cuffs.

In addition to the braided blue patrol jacket, officers were to have, from 1891, a scarlet patrol jacket with five-button fastening, with collar, cuffs and shoulder straps in the facing colour without lace, and two patch pockets on the breast and in the skirts. From 1896 an identical blue jacket, but without facings, replaced the braided type.

The Scottish version of the serge frock had a six-button fastening, rounded skirt fronts, scarlet gauntlet cuffs, and pocket flaps at the waist with three buttons and short loops. Highlanders still had their white shell jackets for drill and barrack wear, but for manoeuvres and active service the frock was worn.

This period saw the last appearance of the red coat on the battlefield. After the Second Afghan War an all-khaki uniform of helmet, frock, trousers and puttees became regulation for all regiments in India in 1885. However, in the Transvaal and Egyptian Wars of 1881 and 1882 the home service undress uniforms were worn with white foreign service helmets, the latter usually dyed. For the Sudan campaigns of 1884–1885 suits of grey serge were issued (Fig. 69, Plate 23). In the later stages of the campaign these were changed for garments of khaki cotton drill in a sandy hue, with which puttees were served out. Nevertheless, the troops had their scarlet serges in their kits and these were worn in action for the last time, as explained in the Introduction, at the Battle of Ginniss in 1885. It was worn in this battle by the 1st Yorkshire Regiment (later Green Howards), 1st Royal Berkshires, 1st Royal West Kents and the Cameron Highlanders. In this action, the frocks were worn over either home service blue or khaki trousers with puttees.

At home the Dorsetshire Regiment was issued with an experimental grey tweed uniform in 1881. This was withdrawn, but three years later a second experimental service dress in 'a warm drab-grey' was produced for troop trials. The frock had pleats like a Norfolk jacket, bronzed buttons, skirt pockets, six removeable cartridge loops on each breast and two upstanding shoulder pieces to prevent straps or rifle sling slipping off the shoulders. The only embellishments were red

1st Battalion Northamptonshire Regiment on manoeuvres, 1892, wearing the experimental white helmet introduced in the late 1880s.

Detachment of 2nd King's Own Yorkshire Light Infantry with Maxim gun in India, 1897. They wear the khaki drill service uniform adopted in India from 1885, Slade-Wallace equipment and are armed with Lee-Metford rifles. The Maxim was issued to battalions from 1890.

piping on collar and cuffs and a red-embroidered regimental title on the shoulder pieces. The shoulder pieces were soon removed as being inconvenient when the greatcoat was worn. Despite the practicality of such a dress, there were strong objections to it, largely on sentimental grounds, and it was abandoned. The coloured serge frocks continued to be worn on all home manoeuvres up to the end of this period.

Overseas, experience in India and the Sudan had proved the advantages of khaki and in 1896 an all-khaki uniform of cotton drill was approved for all foreign service. This was first worn in action during the Sudan campaign of 1898 (Fig. 70, Plate 24), and a year later it was worn on an even wider scale in the South African War — the first major conflict in which a substantial British field army was entirely clothed in khaki. The khaki frocks of Scottish regiments had their usual minor differences and were worn over kilts, sporrans, hose and khaki spats by Highlanders and over trews and puttees by Lowland regiments. The marksmanship of the Boers soon induced the Highlanders to adopt khaki kilt aprons and the Lowland troops to adopt khaki trousers.

Field experience in the war resulted in two other changes. Firstly, the cotton drill of the uniform proved neither hard-wearing nor warm enough and was changed to serge, which, since large quantities had to be produced in a hurry by different contractors, varied considerably in the shade of khaki. Secondly, it soon became apparent to many officers and men that the slouch hat worn by the Colonial troops was a lighter and more practical headgear than the helmet. This began to be adopted on an unofficial basis, some officers approving of it as practical, others vehemently objecting to it as unsoldierly. Nevertheless its advantages were perceived by the home authorities and many of the reinforcing drafts sent out in the latter stages of the war, including those for Guards battalions, were equipped with slouch hats instead of helmets. The foreign-service helmet was in any case coming to the end of its life, as a new pattern with flatter brim called the Wolseley had been designed. This had been worn by a few regimental officers in the Sudan campaign, but its use in South Africa seems to have been largely confined to the Staff, doubtless to avoid regimental officers being picked out from their men by the Boers. For similar reasons, not to mention their uselessness in open warfare, swords were discarded and many officers carried rifles like their men.

Maxim gun detachment, Royal Warwickshire Regiment in South Africa, 1901, wearing slouch hats, khaki serge, and webbing bandoliers.

The 20th Century

BACKGROUND

The basic concept of the 1881 reorganisation for a Line regiment centred on a geographical area finding two Regular and a varying number of reserve, part-time battalions (Militia and Volunteers, later Special Reserve and Territorials). This continued up to and just beyond World War 2. The manpower requirements of World War 1 swelled regiments to an unprecedented number of battalions, most of them raised for the duration of the war only and known as 'Service' battalions. In 1915 the formation of the Welsh Guards completed the national representation in the Brigade of Guards; this was the only new British regiment raised during World War 1. In 1922 five of the Irish regiments were disbanded. In the inter-war years the pre-war organisation was resumed, until 1939 called for another expansion, though not as numerous as that in 1914–1918. The only new regiment formed during World War 2 was the Parachute Regiment but, following the division of the old Indian Army between India and Pakistan in 1948, the British Infantry was further augmented by the 2nd, 6th, 7th and 10th Gurkha Rifles, each of two battalions.

In the same year came the first erosion of the Line Infantry as constituted in 1881 — the reduction of all regiments except the Parachute Regiment and the Gurkhas to one Regular battalion. A few regained their second battalions briefly in the early 1950s but not for long. The decision, in 1958, to revert to an entirely volunteer Army, as opposed to the conscript force instituted in 1939, resulted in the need for major cuts and reorganisation in the Infantry. Between 1958–1970 the 64 one-battalion Line regiments existing after the 1948 reductions (including Rifle Brigade) diminished by amalgamation or disbandment to twenty-eight. Of these, twenty-two are single Regular battalion regiments, the remaining six varying between two and three Regular battalions. These six were all formed from 1964 onwards, by converting single-battalion regiments, several already the product of other amalgamations, into battalions of a larger regiment, rather as occurred in 1881. In the same period the two senior regiments of Guards lost their third battalions, and all the Gurkha regiments, except the senior, lost their second battalions. In 1967 all Guards and Line regiments were grouped into administrative 'divisions'. The details of this process are shown in Appendix 3. Today the Regular strength of the British Infantry is:

Guards	5 regiments, finding 8 battalions
Line	28 regiments, finding 39 battalions
Parachute	1 regiment, finding 3 battalions
Gurkhas	4 regiments, finding 5 battalions

Within the battalion, the experience of the Boer War showed that the old eight-company organisation, each divided into two half-companies, was unwieldy and inappropriate for modern war. By 1914 this had been replaced by a battalion of approximately the same strength, 1,000-odd at war establishment, but divided into four rifle companies, each subdivided into four platoons commanded by subalterns. And so, basically, the battalion has remained, though the number of platoons within a company has varied, and at times the number of rifle companies has been reduced to three. However, developing technology and the complexity of war have required a much larger headquarters and administrative element, with increased numbers of drivers, signallers and other tradesmen, who have been grouped into a wing or company of their own. Furthermore the need for infantrymen to have their own heavy weapons, machine-guns, mortars and anti-tank weapons, has seen a reduction in men armed with rifle and bayonet in favour of specialists trained to crew such weapons. The latter have generally been grouped together into a separate support company for administrative

Privates, 1st Battalion Scots Guards, 1903, in 1902 service dress with Brodrick cap and slouch hat. The right-hand figure is in fatigue dress. Leather gaiters were still worn, pending the issue of puttees.

purposes, but decentralised to rifle companies in action. A summary of such weapons since the introduction of the Maxim machine-gun in 1890 is given in Appendix 4.

Ever since the Boer War, infantry tactics have been based essentially on a combination of fire and movement, with particular emphasis on the use of extended order, a high standard of marksmanship, the skilful use of ground and deployment in depth, both in attack and defence. The admirable tactical standards attained by the pre-1914 Regular infantry lapsed during the war, when the need to train mass armies quickly, and the static conditions of the Western Front, led to a return of the more rigid and formalised manoeuvres, and formations, reminiscent of an earlier age. The advent of heavy artillery and the armour of the tank brought much greater firepower to support and cover the infantry's movement, but the huge casualties resulting from inflexible tactics used against strongly defended machine-guns, sited in depth, drove home the lesson, practised in the latter stages of the Great War and afterwards, that infantry should work in small groups, each supporting the other.

The disciplined and accurate musketry, for so many years the great strength of the British Infantry, whether with musket or rifle, now yielded place to the firepower of light and medium machine-guns, mortars and later, anti-tank weapons. The role of the rifleman, though ultimately the man who would actually close with the enemy, became more the protector of his supporting weapons, which in attack helped the movement forward and in defence provided the framework of the position held.

The need for the closest co-operation on the battlefield between armour, artillery, engineers and infantry became evident during World War 2. Since then, the advances in lethality and capability of modern tanks and guns, not to mention the threat of tactical nuclear weapons, have not only led to the mechanisation of the infantry, but also to the disappearance from the battlefield of separate armoured regiments and infantry battalions in favour of the battle group — a flexible grouping in which, at any one time armour or infantry may predominate according to the circumstances. The battalion remains the basic infantry unit for peacetime administration and for counter-insurgency operations, but on the nuclear or conventional battlefield a number of its companies will form, with one or more armoured squadrons, an armoured or mechanised infantry battle group.

DRESS 1902–1918

The lessons of the Boer War relegated the full dress, of scarlet or rifle-green to ceremonial purposes and walking-out. This dress remained much as in the previous period, except that from 1902 the cuffs were again pointed, the back skirts acquired slashed flaps with buttons, and in 1913 the shoulder straps reverted to the facing colour. The shoulder titles changed in 1908 from white embroidery to brass. Early in this century the civilian practice of pressing trousers with creases fore-and-aft was applied to military legwear.

In 1903 the Royal Scots and King's Own Scottish Borderers exchanged their unsuitable helmets for broad-topped blue Kilmarnock caps with diced bands, red touries on top and black-

Royal Welch Fusiliers, 1904. Corporal-drummer in full dress, lance-corporal in service dress with 1903 Bandolier equipment and short magazine Lee-Enfield rifle.

cock feathers on the left side behind the badge. No changes occurred to the full dress headgear of other regiments, but in 1902 the new Brodrick cap replaced the field service cap for undress and walking-out. The Brodrick cap was round, dark blue, with a flat top, and the badge in front. Never popular, it was changed in 1905 for a forage cap of similar shape and colour, but with a wired brim, a drooping black leather peak and scarlet band for Royal regiments. It had a black patent leather chinstrap fastened by small side buttons, and the regimental badge in front. The Glengarry continued to be worn by Scottish regiments.

The varying quantities of gold lace and braid distinguishing rank on the collar and cuffs of officers' tunics was reduced in 1902 to that formerly worn by subalterns. In the same year the crimson sash was moved to the waist, the sword slings of red leather edged with gold lace being suspended from a belt worn under the tunic. Scottish regiments continued their previous system of belts with the sash still worn over the left shoulder. On the outbreak of war in 1914 all full dress was withdrawn.

In 1902 the former undress uniform was abolished and replaced for all occasions, except those requiring full dress, by a khaki service dress. This was to be worn either with the Brodrick or Glengarry cap, or when on manoeuvres, with a slouch hat turned up on one side with the regimental badge and, in the case of the Guards and Fusiliers entitled to plumes in their full-dress headgear, similar plumes but shorter. From 1905 a round peaked cap of khaki serge became the universal service headdress, except for the Glengarries of Scottish regiments.

The tunic of the new dress was of 'drab-mixture' serge, a shade with more green in it than the khaki drill. It had a turned-down collar, reinforcing patches at the shoulders, two patch breast pockets with pleats, two side pockets with flaps, and detachable shoulder straps piped red. All buttons were of gilding metal, 'which, when not polished for some time, assumes a dull colour matching the material'. Rifle regiments had black buttons. Collar badges were not worn but regimental titles or initials were embroidered in white on a red background on a curved strip of material sewn just below the shoulder seam. The number of the battalion was on a separate patch below the title (see Fig. 73, plate 25). In 1908 these embroidered titles were replaced by brass titles on the now fixed shoulder straps, from which the piping

Officer, Rifle Brigade, full dress, 1904. In his right hand is the black lambskin busby with black plume introduced in 1890.

Piper, Scots Guards, and drummer, Irish Guards, full dress, 1910. The piper's doublet is dark blue, the tartan Royal Stewart. Watercolour by P. W. Reynolds.

Plate 25: 1902–1918

73. Lance-Corporal, 2nd Battalion, The Buffs (East Kent Regiment), 1903. 74. Private, 1st Battalion, Seaforth Highlanders (Ross-shire Buffs, Duke of Albany's), 1908 (India). 75. Drummer, South Wales Borderers, 1913. Fig. 73 shows the serge service dress, in darker khaki, introduced for all purposes in temperate climates, except ceremonial and walking-out. Initially it had embroidered regimental shoulder titles and the battalion number underneath. Puttees were prescribed for this dress but leather gaiters continued in use until stocks ran out. In barracks the peakless Brodrick cap was worn; in the field the slouch hat. Both of these were replaced by the khaki peaked cap in 1905 (see Fig. 81, plate 27). Equipment was the modified Slade-Wallace with haversack on the back until the Bandolier pattern (Fig. 74) appeared in 1903. Fig. 74 shows a Highlander on the Mohmand Expedition of 1908 wearing the new Wolseley helmet with quilted curtain, khaki drill service dress tunic, kilt apron and khaki spats. A mess-tin in khaki cover and rolled ground sheet is on the back of the belt. At home the Bandolier equipment was completed by the greatcoat folded across the shoulders. Full dress (Fig. 75) continued to be worn until 1914. The round cuffs ordered in 1881 reverted to pointed cuffs in 1902. This regiment was one which regained its pre 1881 facings instead of its national colour. All Line drummers' tunics had a universal lace of red crowns on white.

73. L/Cpl, 2nd Buffs, 1903.

75. Dmr, S. Wales Borderers, 1913.

74. Pte, 1st Seaforth Hldrs, 1908.

76. Offr, 1st Cameronians, 1914.

78. Pte, 1/22nd London Regt, 1918.

77. Pte, 1st Leinster Regt, 1916.

Drummer, Black Watch, drill order, and sergeant, Rifle Brigade, review order, c 1905. The sergeant's rifle, a Lee-Enfield, has the sling loose, according to the custom of Rifle regiments. Watercolour by P. W. Reynolds.

Plate 26: 1902–1918

76. Officer, 1st Battalion, The Cameronians (Scottish Rifles), 1914 (France). 77. Private, 1st Battalion, The Prince of Wales's Leinster Regiment (Royal Canadians), 1916 (Near East). 78. Private, 1st/22nd Battalion, The London Regiment (T.F.), 1918 (France). This plate shows three figures from World War 1 in the 1902 service dress worn throughout. In 1914 all Scottish regiments wore the coloured Glengarry, later changing to the khaki Balmoral bonnet. The officers' tunic was worn with a collar and tie; the rank badges being placed on the cuffs. Tunics of Scottish regiments had cut-away skirts and their officers had gauntlet cuffs instead of the slashed pattern of other regiments. Fig. 76 wears Sam Browne equipment with haversack, revolver, binocular and compass cases. All officers, except those of kilted regiments, had breeches worn with puttees or leather gaiters for mounted officers. All officers carried swords early in the war but soon discarded them. Fig. 77 (of an Irish regiment) shows the mixed serge and drill costume adopted in the Near East, worn with Wolseley helmet and the 1908 webbing equipment; shorts had been introduced in India before the war. Fig. 78, of a Territorial Force battalion, wears the steel helmet issued in 1916 and a leather jerkin better suited to the trenches than the long greatcoat. He carries the automatic Lewis gun as its No. 1 and is armed with a revolver. He has a modified set of the 1914 leather equipment. The entrenching tool handle hangs from his bayonet frog.

had been removed. Guards retained the embroidered type. The khaki trousers were to be worn with puttees, but the old black gaiters continued in use for some years until stocks were reduced.

The tunics of Scottish regiments had their usual distinctive features of cut-away skirts and gauntlet cuffs. Highlanders wore kilts and hose with khaki aprons and spats but Lowland regiments had khaki trousers and puttees.

The Slade-Wallace equipment continued for a year or two, sometimes with the pouches exchanged for webbing bandoliers or the fifty-round leather type with flaps, used by cavalry — a practice widely adopted in South Africa. In 1903 a universal Bandolier equipment replaced the Slade-Wallace for wear with service dress. However the latter's waistbelt and frog was retained by the Line for full dress, and the Guards, from 1905, wore a modified Slade-Wallace equipment in full dress guard order, dispensing with haversack, water bottle and valise, but with the cape rolled at the back of the waistbelt and the folded greatcoat taking the place of the valise.

The Bandolier equipment consisted of: a brown

Gloucestershire Regiment, 1908. Private, review order, sergeant, Indian khaki drill with Wolseley helmet and Bandolier equipment. Note that the shorts, adopted about this time, practically covered the knee. Watercolour by A. C. Lovett.

leather waistbelt with square brass buckle, to which were attached two brown leather pouches containing fifteen rounds in five-round chargers, two more with ten rounds, and a fifth pouch holding a tin of rifle-oil; the bandolier itself (with five 10-round pouches) over the left shoulder and secured on the right side by a buckle and strap to the waistbelt; the folded greatcoat (khaki instead of grey from 1904) secured high on the shoulders in a webbing harness with braces which clipped to the fifteen-round pouches on the belt and a strap attached to a **D** on the canvas cover of the mess-tin, which itself was looped on to the waistbelt at the rear; haversack and water bottle, suspended by separate slings over the right and left shoulders respectively.

This was the last of a long line of leather accoutrements, for in 1908 a new equipment, made entirely of webbing and of revolutionary design, was approved. For the first time all the components fitted together so that the whole assembly could be put on or taken off like a coat. It was based on a 3 in. wide waistbelt with brass **D** s at the back, and two 2 in. wide braces passing over the shoulders and crossing at the back, to which the pack was buckled. On either side of the expanding belt fastening were five 15-round pouches, three on the belt and two above, attached to the braces. The free ends of the braces were buckled at front and back to the haversack on the left side, and water bottle and entrenching tool carrier on the right.

To complement the new uniform and equipment, a new rifle was introduced, the short Lee-Enfield. To compensate for its reduced length in close quarter fighting, the bayonet blade was increased to 17 in.

A broadly similar khaki service dress, but of finer material, was introduced for officers at the same time as the men's, except that breeches with puttees or leather Stohwasser leggings for mounted officers, were worn instead of trousers. At first, rank was distinguished by a graduated arrangement of khaki braiding on the cuffs. This was soon displaced by a cuff slash outlined in drab lace, wherein were placed worsted rank badges of the same combination of crowns and stars as worn on the full dress tunic, with rings of drab chevron lace and tracing braid, varying according to rank, passing round the cuff from the centre point of the slash. Scottish regiments had no slash, but a similar arrangement on a gauntlet cuff. Guards officers wore their rank badges on the shoulder straps, had no pleats in the breast pockets and had their buttons grouped according to regiment. From 1908 all officers abandoned the turned-down collar, and, instead, had an open step collar showing the khaki shirt and tie. In most Line regiments the cap and collar badges were of bronzed metal.

The Sam Browne belt was worn with this uniform. It was worn on normal duties with single cross-brace and on field service with double braces, sword frog, revolver holster, pouch, haversack and water bottle; the latter had a separate strap but the haversack was clipped to the belt on the left side.

Besides their full and service dress, officers still had an undress uniform, which saw a return to the mid nineteenth century in the shape of a double-breasted blue frock coat with collar badges and gilt buttons. This was worn with a dark-blue peaked forage cap, similar to that introduced for the men in 1905. The frock coat was authorised for Lowland regiments but Highland officers assumed a white shell jacket, of the type worn by their men for many years. Guards officers had their own pattern of braided frock coat which had remained largely unchanged since the 1830s.

In addition to this undress uniform, officers also had mess dress, which had developed in the previous century from the custom of wearing the shell jacket at mess; this garment had provided the basis for the first universal mess jacket authorised in 1872. From 1896 a jacket with roll collar, worn open to reveal the white shirt, black bow tie and waistcoat had been authorised, but some regiments kept to the old pattern with standing closed collar.

In tropical climates a khaki drill version of the new service dress was worn with the Wolseley helmet mentioned in the last period. This helmet was covered in khaki cloth, thus dispensing with the need for the khaki covers worn over the men's white helmets of the previous pattern. For these helmets, regiments devised their own distinguishing marks, usually sewn to the puggaree on one side. From 1905, earlier in some regiments, the practice of wearing shorts with puttees up to the knee became popular. On ceremonial occasions in India, khaki gave way, in the cold weather, to a scarlet serge frock similar to that worn in the 1870s, blue trousers, and white Wolseley helmet with puggaree badge; in the hot season, to an all-white uniform.

It was in the 1902 service dress, or its tropical

NCOs and men, full dress, 1914. From left: Royal Fusiliers; K.R.R.C; Line corporal and Lancashire Fusiliers (*at back*); Lincolnshire Regiment; Rifle Brigade; Line (*pointing at back*); King's Own Royal Regiment; K.O.S.B. (*at back*); Coldstream Guards; Northumberland Fusiliers (*back*); Seaforth Highlanders; A.S.C.; Black Watch (*back*); R.E.; Cameronians (*back*); Royal Scots Fusiliers; Military Police; Royal Dublin Fusiliers; R.A.; A.O.C. After a watercolour by W. McNeill.

equivalent, with the 1908 equipment that the Infantry went through World War 1. After the opening months men started to remove the stiffening wire which gave their caps a smart appearance. Later, a version with folding ear flaps, soft top and peak was used. In early 1915, Scottish regiments abandoned their colourful Glengarries in favour of a flat khaki bonnet with tourie, called a 'tam o' shanter' or Balmoral. In 1916 the steel helmet was issued.

After a time, the tunic pockets were made without pleats as an economy measure, and the sleeves began to acquire a variety of coloured formation signs, wound stripes and service chevrons. Some regiments converted their trousers into shorts for summer wear. Highlanders kept their kilts throughout the war but with the apron all round, and the diced hose and spats were exchanged for khaki hose-tops and short puttees.

In 1914 officers took their swords to the front but these were soon discarded. Furthermore the casualty rate among company officers forced them

Grenadier Guards, 1914. From left: Company-sergeant-major, service dress, marching order with 1908 equipment; musician, full dress; officer, service dress, drill order; private, full dress, marching order. Watercolour by D. Macpherson.

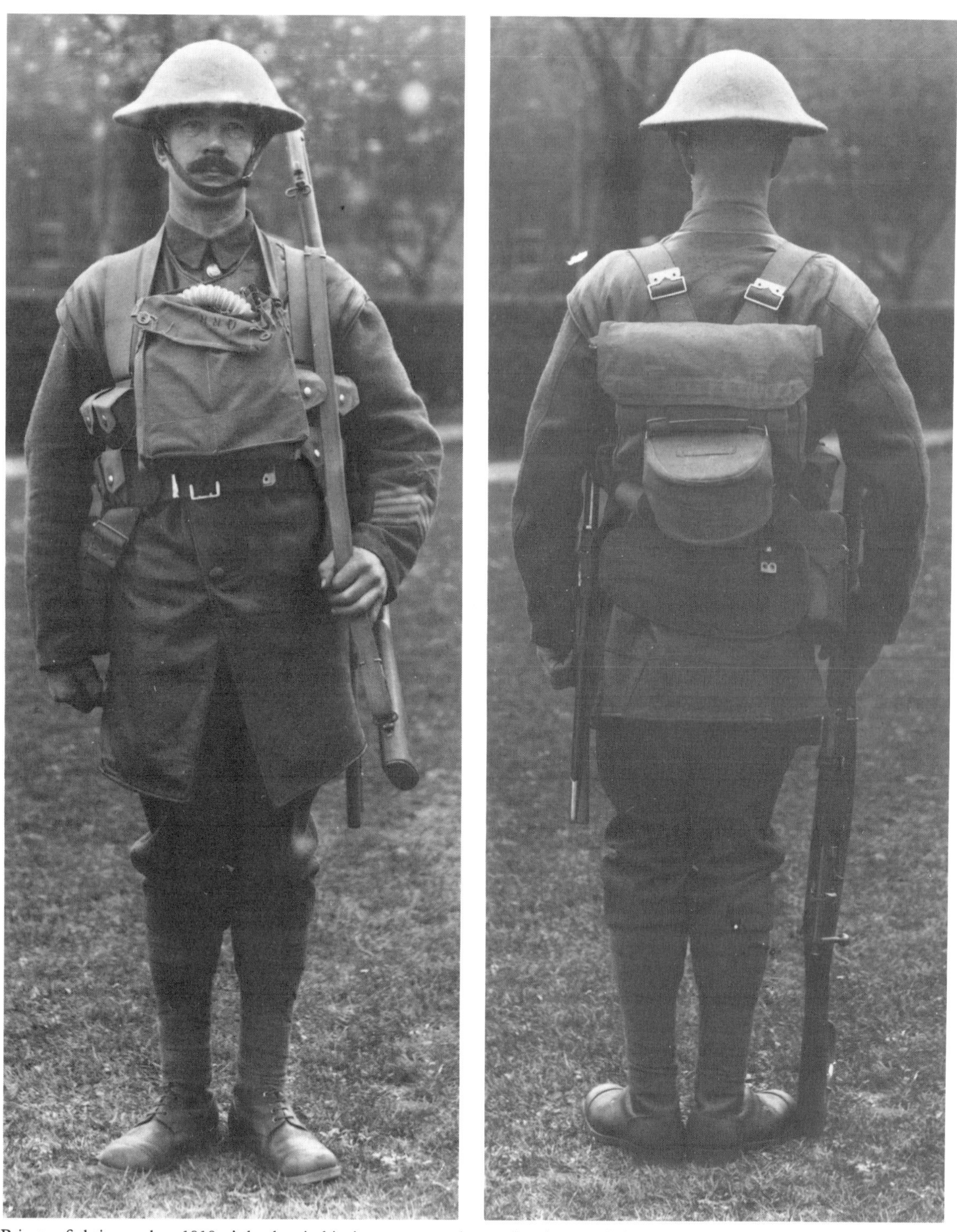

Private, fighting order, 1918. A leather jerkin is worn over the service dress and the steel helmet is covered in sacking. The haversack of the 1908 equipment is worn as a pack with mess-tin attached. The entrenching tool carrier hangs below, its helve strapped to the bayonet scabbard. The box respirator is at the 'Ready' position.

to assimilate their appearance to the men's, adopting rifles and soldier's equipment, even their tunics. Their rank badges, being too conspicuous on the cuffs, were moved to the shoulder straps.

In the winter months the long greatcoats proved inconvenient in the muddy, water-logged trenches and shorter, goatskin coats were issued instead. Later in the war sleeveless leather jerkins were served out.

In the open warfare of 1914, the full web equipment was carried but once operations became static the large packs were usually left off and the haversack took its place on the shoulders. The stocks of webbing equipment were insufficient to supply all the newly raised Service battalions and, in 1914, a substitute equipment was manufactured. This followed the principle of the 1908 pattern, but the belt, braces, water-bottle carrier and twin pouches (the latter similar to the Slade-Wallace type) were of brown leather. The pack and haversack were of webbing but fitted with leather straps. The introduction of poison gas as a war weapon necessitated the issue of a respirator. The first type was carried in a small satchel slung over one shoulder, but later on the box pattern appeared, which was worn on the chest.

In the war outside Europe a variety of dress was worn, this included the normal home-service serge, either with cap (sometimes fitted with neck curtain) or Wolseley helmet, khaki drill with shorts, or a mixture of serge tunic and khaki drill shorts. The Wolseley helmet was the most common headgear in the Near and Middle East campaigns, though the slouch hat made a reappearance and the steel helmet was worn in the final operations in Palestine, sometimes fitted with a neck curtain.

DRESS 1919–1939

After World War 1, full dress, though not abolished, was not re-issued to the bulk of the Army and its use was confined to the Brigade of Guards, officers attending Levées, and bands, corps of drums, bugles and pipes on special occasions. On these latter occasions, such as the Aldershot Tattoos of the 1930s, the dark blue forage cap was more customarily worn than the spiked helmet.

Only two new full-dress uniforms need to be noted in this period. First, that of the Welsh Guards, formed in 1915, who had their buttons in fives and a white/green/white plume on the left side of the bearskin. Second, a novel costume appeared in Irish regiments. After the disbandment of five Irish regiments in 1922, the national character of the remaining four was emphasised by the addition of pipers, clad in a supposedly Irish dress of 'caubeen' bonnet with hackle plume, doublet, saffron kilt, hose and cloak. In full dress the pipers of the Irish Guards had these garments, except the kilt, in green; the plume was blue. Drummers of Irish pipe bands were clothed as the rest of the regiment.

Although Guards regiments reverted to the pre-war arrangement of full dress for ceremonial occasions and service dress for everything else, Line regiments only had service dress. It was, however, considerably smartened up. Caps regained their stiffened crowns, tunics, with collar

Plate 27: 1919–1939

79. Officer, The Queen's Own Royal West Kent Regiment, 1930. 80. Guardsman, Irish Guards, 1932. 81. Corporal, The Gloucestershire Regiment, 1933. This and the next plate show dress at home and abroad in the inter-war years. Fig. 79 is an officer dressed for manoeuvres at home in Sam Browne equipment, with anti-gas respirator on his chest. The main change from pre-war service dress is the use of 'plus-fours' type of trousers. He is armed with revolver and carries a walking stick. Only Guards regiments now retained full dress, which in essentials had changed little over the previous sixty years. The Irish Guards' buttons in fours, and the Welsh Guards' buttons in fives, continued the three senior regiments' practice of grouping the buttons. Fig. 80 is in 'Guard Order', with Slade-Wallace equipment, greatcoat folded flat across the shoulders and attached to the braces where the valise was formerly carried, and cape rolled at the back of the belt. From 1936 the greatcoat was removed, the folded cape replacing it. Fig. 81 shows the smartened up service dress as worn on ceremonial occasions in the 1930s by Line regiments. Skill-at-arms badges are sewn on the left forearm. Only the waistbelt and frog of the 1908 equipment was worn in this order of dress. The 'back badge' at the rear of the cap commemorates the Battle of Alexandria, 1801.

79. Offr, Royal West Kent Regt, 1930.

81. Cpl, Gloucester Regt, 1933.

80. Gdsmn, Irish Guards, 1932.

82. Sgt, Grenadier Guards, 1930.

84. Pte, Hampshire Regt, 1939.

83. Offr, 1st Northampton Regt, 1937.

All ranks, Scots Guards, c 1930. From left: two officers, blue patrol; officer, frock coat; company-sergeant-major, guard order; officer, service dress; regimental-sergeant-major, full dress drill order; officer, guard order; piper, full dress; drill sergeant, guard order; orderly, service dress; drum-major, state dress; guardsman, drill order; guardsman, fighting order; drummer, full dress; guardsman, guard order.

Plate 28: 1919–1939

82. Sergeant, 1st Battalion, Grenadier Guards, 1930 (Egypt). 83. Officer, 1st Battalion, The Northamptonshire Regiment, 1937 (India). 84. Private, The Hampshire Regiment, 1939. When abroad, khaki drill and Wolseley helmets were worn for all purposes. Trousers were customary for ceremonial; shorts, puttees and hose-tops for other duties. Fig. 82, with regimental white plume in the puggaree, wears the Slade-Wallace belt and one pouch. Regiments in India in a similar order of dress to Fig. 82 used the brown leather belt of the 1903 equipment. Fig. 83 is dressed for the Waziristan operations of 1936–1937 on the North-West Frontier, in which all ranks wore the lighter, pith topee instead of the Wolseley helmet; a black regimental flash is stitched to the side. A jersey and shirt replaced the KD tunic, and short puttees were used instead of long puttees. Coloured flashes on the shoulder straps denote the wearer's company. Officers wore the jersey outside the shorts, soldiers inside. The Sam Browne equipment had given way to the 1937 pattern for officers, with revolver holster and pouches for ammunition, binoculars and compass, all in webbing. A haversack on the left side balanced the water bottle on the right. The full 1937 equipment for soldiers appears in Fig. 84 with the new 'battledress', devised in 1938 to replace the 1902 service dress. The field service cap, first worn in the 1890s, was re-introduced in khaki for this costume. He is equipped with the Bren light machine gun which replaced the Lewis gun.

badges added, were better tailored and more close-fitting, trousers were smartly cut and pressed into pleats where they turned over the tops of the puttees. Unofficially the same effect was sometimes achieved by cutting the trousers short just below the top spiral of the puttee. On parade occasions Scottish regiments wore trews or kilts with Glengarrries, though retaining the tam o' shanter for training.

When in service dress, Guards wore either a khaki cap like the Line's cap, but with a shorter peak, or their coloured forage cap with distinguishing band — Grenadiers red, Coldstream white, Scots diced, Irish green and Welsh black. The equivalent caps for Guards officers had black bands in all regiments except the Scots Guards.

In order to preserve their serge uniforms, all soldiers were issued with a suit of brown canvas 'fatigues'. These were worn for the duties after which they were named, and sometimes at drill or for training in barracks.

The 1908 equipment remained in use through most of this period, its belt and bayonet frog being the only items of the equipment worn on parade or ceremonial. Some regiments had the Slade-Wallace belt for special occasions or when walking-out. In full dress guard order, the Foot Guards still used the modified Slade-Wallace equipment as in pre-war days, but from 1936 the cape, formerly rolled at the back of the belt, took the place previously occupied by the greatcoat, which was discarded in this order of dress.

Officers' service dress was much as before, except that the wartime practice of wearing rank badges on the shoulder straps was now regularised. Mounted ranks, i.e. field officers, company commanders, adjutants and transport officers, continued to wear breeches with either knee boots or leather leggings, but other officers adopted rather baggy trousers, similar to civilian 'plus-fours', worn over puttees. The undress blue frock coat was seen no more, except in the Foot Guards, but all officers still had their scarlet mess jackets and a blue patrol jacket of 1896 pattern, both worn with blue overalls (close-fitting trousers strapped under the boot) and either the 1905 coloured forage cap or the field service cap.

Soldiers' tropical, khaki drill uniform was the Wolseley helmet, tunic, and either trousers without puttees, or shorts, long puttees and woollen hose-tops turned over the top spiral. Trousers were customary for ceremonial; shorts for daily wear and training. In India, soldiers were issued with a brown leather waistbelt of the 1903 equipment for wear when walking-out and for parade purposes. On ceremonial occasions two brown leather pouches from the same equipment were added to the waistbelt.

In certain foreign stations, e.g. Shanghai, Palestine, and Egypt, home service clothing with caps was worn in the winter months. In India, although serge tunics were sometimes ordered in cold weather, the helmet was worn all the year round. In India, in the late 1930s, the Wolseley pattern was reserved for parade and ceremonial; for training, or operations on the North-West Frontier, a light flat-topped pith topee was issued (see Fig. 83, Plate 28). The KD tunic, too, was kept for more formal occasions, soldiers in the field or on manoeuvres wearing shirtsleeve order, supplemented by the woollen khaki jersey issued at home, when necessary. Short puttees were also introduced for wear with hose-tops.

An exception to the universal wearing of service dress was made for troops taking part in the Coronation of King George VI in 1937. Since the war, soldiers had been permitted to purchase a blue patrol jacket and trousers of the full-dress pattern for wear when walking-out. If worn, the headdress was to be the coloured, peaked forage cap. It was decided to issue a dress of this nature to the Coronation contingents, except of course to the Foot Guards who already had full dress. For Rifles, it was rifle-green, Light Infantry had dark green forage caps, and Scottish regiments wore Glengarries with either trews, or kilts, sporrans, hose and spats; the jackets of the Scottish regiments had cut-away skirts. Though it lacked the splendour of full dress, this all-blue uniform gave a more ceremonial touch to an important occasion than utilitarian khaki would have done. The effect was somewhat marred, however, by the wearing of khaki web belts. Officers paraded in their normal patrol dress with Sam Browne belts.

Despite the general loss of full dress, the Army's clothing in the 1920s and 1930s was of a smart and soldierly appearance and, furthermore, had stood the test of the greatest war known to man. Nevertheless, in the 1930s, revolutionary changes were afoot. In 1933 an experimental new service dress was produced for trials. This had a novel headgear derived from the deer-stalker hat. The upper garment was still a skirted tunic type, with four pockets, but with closely buttoned cuffs and a collar which could be worn open at the neck. Legwear was either shorts with puttees or

Private, Argyll and Sutherland Highlanders, c 1930. Left: field service marching order with tam o' shanter, or Bonnet TOS. The large pack is on the back with respirator on top, its sling secured under the bottom tunic button. Right: fighting order. The haversack has replaced the pack. A steel helmet and kilt apron are worn.

trousers with webbing gaiters. Having undergone tests, this clothing was shelved.

Then, in 1938, the War Office signalled the end of the tunic era by approving a uniform which completely broke away from anything seen before, and resembled a suit of workman's overalls — battledress. Ignoring the need for a soldier's upper garment to afford some covering below the waist, so long urged in the mid nineteenth century, the new dress had a capacious blouse which stopped short at the waist. It had two breast pockets with concealed buttons, a front opening like a trouser fly, and it was fastened at the right front of the waist by a buckle and a tongue of material. At the back of the waistband were two buttonholes for securing the blouse to two buttons sewn on the back of the trousers. The trousers were generously cut, had one hip and two side pockets, an open pleated pocket in front of the right hip for the first field dressing and a large deep pocket on the left thigh. At the base of each trouser leg was a short flap and button to close the trousers round the ankles, although the authorities also intended to issue short webbing gaiters, or anklets, to protect the gap between trouser bottoms and boot tops. As a headdress for this shapeless costume, when the steel helmet was not worn, the field service cap of the 1890s was resurrected but in khaki. Except for the cap badge and two small brass buttons (black for Rifles) on the front, all metal insignia and buttons were done away with, although regiments were to be permitted khaki slip-on shoulder titles with their designations worked in black. Guards and Scottish regiments were excused the field service cap, retaining their peaked khaki caps or tam o' shanters respectively.

A new equipment, the 1937 Braithwaite pattern, was also introduced. The principles of the 1908 type were still adhered to, the pack and water-bottle carrier were unchanged, but the haversack was a little larger, the belt and braces narrower, and the ten individual pouches gave way to two large deep ones. The latter were necessary to accommodate the long curved magazines of the Bren light machine-gun, then coming into service as a section weapon to replace the Lewis gun. To ensure an adequate ammunition supply for the Bren, each man of a section would have to carry a proportion of its rounds. In full marching order the whole equipment would be worn, but in battle order the haversack would replace the pack on the man's back, so that only the bayonet and water bottle, if not carried in the haversack, would hang below the waist.

Officers were to wear a battledress identical to the men, though collars and ties were permitted, with the blouse unfastened at the neck. For the first time they too received a web equipment of the same Braithwaite pattern, but instead of the so-called 'basic' pouches, they had a revolver holster, two small pouches for ammunition and compass, and a larger rigid box-like pouch for binoculars.

Battledress was to be universal for all ranks and all regiments; even Highlanders had to forego their kilts and hose. So, for the first time in the history of uniform, everyone, commissioned or otherwise, Guards, Line, Fusiliers, Light Infantry, Scottish or Rifles, was to be dressed uniformly, or as near uniformly as the British military tendency towards diversity in such matters would permit. This was doubtless a praiseworthy intention. Unfortunately, it was to prove not an

Officer, NCOs and privates, 1st Battalion Northamptonshire Regiment, in khaki drill, Egypt 1931. The men wear home service puttees with hose-tops, the officer's puttees are a lighter shade. Note the black regimental flash on the side of the helmets.

ideal dress for, apart from its unsightliness, the cap tended to fall or blow off, the blouse to part company with the trousers, exposing the midriff and the trouser bottoms tended to pull out of the anklets. As will be seen, however, it was to last for nearly a quarter of a century.

Battledress and the new equipment began to be issued in 1939 but, when war broke out in September, battledress had by no means reached all regiments, and in the expeditionary force sent to France the old uniform was still to be seen, sometimes mixed with items from the new.

DRESS 1940–1962

This period embraces not only World War 2, but also the post-war years up to the ending of National Service, or conscription. This era of the Army's history coincides with what may be called, from a dress viewpoint, the battledress period.

Full dress naturally disappeared during the war, from those entitled to wear it, but was re-introduced in 1948 for the Brigade of Guards when performing public duties. The Guards' basic uniform remained as pre-war, but the modified Slade-Wallace equipment was further decreased, leaving the guardsman with only waist-belt and bayonet frog. Before the war the tunics of Guards' musicians had had gold-laced wings and loops on the chest, in the same arrangement as the drummers' fleur-de-lys pattern lace, but in 1948, although drummers' tunics were unchanged, musicians had to wear ordinary guardsmen's tunics; their wings, however, were later restored. After the war some Line regiments managed to equip their corps of drums, or equivalent, with full dress for special occasions, though with forage caps rather than helmets.

As stocks of battledress increased, so the old service dress disappeared, although officers still wore it when not on parade with troops. The khaki field service cap continued in use until 1943, when an even more unsightly headdress was approved — the cap, general service. This was a round pancake-like object with a browband, not unlike the Scottish tam o' shanter without a tourie, but with the crown pulled over to the right instead of forward. Guards and Scottish regiments were spared this uncomfortable and hideous bonnet, as were the two old Rifle regiments, K.R.R.C. and Rifle Brigade, who as motorised infantry adopted a khaki beret — a headdress of French origin first introduced into the British Army in 1924 by the Royal Tank Corps, who were granted a black beret as more appropriate than the khaki cap. Despite lacking the smartness and uniform appearance of a cap, the beret, as issued, was convenient for wartime conditions and, being made of superior material with a narrow leather browband, made a far better showing than the unfortunate 'cap g.s.'. Many Line regiment officers acquired them and the newly-formed Parachute Regiment were granted one in a distinctive maroon.

After the war it was decided to make the beret the universal headdress for wear with battledress; in dark blue for most regiments, khaki for Guards (when not wearing coloured or khaki forage caps), dark and rifle green for Light Infantry and Rifles respectively. Fusilier regiments were allowed hackles in the colour of their former full dress plumes. Scottish regiments continued to wear the tam o' shanter or the Glengarry. The caubeen with hackle was adopted for Irish regiments (except the Irish Guards) in dark blue for the Royal Irish and Royal Inniskilling Fusiliers, and rifle-green for the Royal Ulster (formerly Irish) Rifles; the hackles were green, grey and rifle-green respectively. Regimental badges were worn above the left eye with caps g.s., berets and caubeens, and on the left side of the tam o' shanter, except for the Black Watch who wore their traditional scarlet hackle.

The battledress blouse underwent a number of changes. As a result of wartime economies the pocket pleats disappeared and the removal of the fly front revealed plain bone buttons. This gave Rifle regiments the opportunity to display their black buttons. After the war, the fly front and pocket pleats returned and, as had happened to the serge tunic after 1918, the blouse became more tailored, in an attempt to give it a smarter appearance, with the multiple pleats at the waistband replaced by two box pleats at the back. Coloured formation signs appeared on the sleeves during the war and continued to be worn afterwards. Arm of service flashes, a narrow strip of cloth, red for Infantry, rifle-green for Rifles, were worn under the formation signs. In the later years of the war coloured regimental titles, similar to those worn on the first issues of the 1902 service dress, were introduced. These were generally white on red for Line regiments except Scottish, who had a piece of regimental tartan. The K.R.R.C. had red on green, the Rifle Brigade and Royal Ulster Rifles black on green. Guards had their own pattern with superior lettering. The Grenadiers and Coldstream had white on red, Scots yellow on blue, Irish white on green, Welsh white on black. In the post-war era several Line regiments adopted titles in their own colours. The Royal Norfolk, for example, had black on yellow,

Grenadier Guards, World War 2. From left: guardsman, battle order, 1944–1945; officer, armoured battalion; guardsman, battle order, 1940–1941; guardsman, Western Desert, 1941 (note Bren LMG in front); drummer, training battalion. After a watercolour by A. E. Haswell Miller.

Men of the Middlesex Regiment in khaki drill and 1937 pattern equipment with Vickers medium machine guns in the Western Desert, 1942. The Middlesex was one of several machine gun battalions. Each Vickers had a crew of three: No. 1 (corporal) and 2 on the gun, No. 3 re-supplying ammunition. Two guns formed a section under a sergeant.

the Royal Hampshires yellow on black. The South Staffords had yellow on maroon, with a glider below commemorating their wartime service as air-landing troops, while the Light Infantry had white or yellow on green.

In 1945 soldiers were allowed to wear the neck of the blouse open to show a collar and tie when walking-out. Later this concession was extended to everyday use and the inside of the collar was faced with serge material. Eventually the neck was made with a neat, open step collar, as on an officer's service dress.

The trousers also underwent modifications. The fastening at the bottoms was removed and, in the early 1950s a second hip pocket was added while the thigh pocket was placed on the outside of the leg to facilitate a better crease in the trousers. Due to their width, it was never possible to achieve as neat an overhang of trousers over anklets as had been possible with the pre-war trousers and puttees. To remedy this, many soldiers unofficially took to wearing weights in their trousers to make them hang better. This gave way to another unauthorised device, which also prevented the trousers pulling out of the anklets, the placing of elastic bands round the anklets and the tucking of the trouser ends up inside them. In regiments which took turn-out seriously, such practices were discouraged; neither were ever permitted in the Foot Guards.

With the introduction of battledress, the old brown canvas 'fatigues' were replaced by a suit of 'denims', on the lines of the battledress but even more shapeless and named from the material of which they were made. These were worn for all

Plate 29: 1940–1962

85. Lance-Corporal, 1st/6th (T.A.) Battalion, The Queen's Royal Regiment (West Surrey), 1942 (North Africa). 86. Officer, 2nd Battalion, The East Yorkshire Regiment (Duke of York's Own), 1944 (North-West Europe). 87. Private, The Parachute Regiment, 1944 (North-West Europe). This plate shows types of World War 2. In the North African and Italian campaigns battledress was worn in cold weather, KD shirts and shorts or trousers at other times. Jerseys, as in Fig. 85, provided an intermediate degree of warmth. This NCO, of a Territorial battalion, has the 1937 equipment and carries a Thompson sub-machine gun, later replaced by the Sten. Both he and the officer in Fig. 86 wear the 1916 steel helmet, superseded in 1944 by the deeper type still worn today. During the war the battledress sleeves acquired formation signs, first worn in World War 1, arm-of-service flashes, and eventually coloured regimental titles as worn in 1902. A new type of infantry, The Parachute Regiment, raised in 1942, wore a distinctive maroon beret, or a special type of steel helmet, and the Denison smock of camouflaged material. Fig. 87 wears 'battle order', in which the haversack of the 1937 equipment became a small pack. The entrenching tool with case hangs from the back of the waistbelt. He is armed with the wartime No.4 rifle and spike bayonet.

85. L/Cpl, 1/6th Queen's, 1942.

87. Pte, Parachute Regt, 1944.

86. Offr, 2nd East Yorkshire Regt, 1944.

88. Pte, 1st Lancashire Fus, 1944.

90. Rfn, 2/2nd Goorkhas, 1960.

89. Pte, 1st D.W.R., 1953.

dirty work, sometimes on operations during the war and for most training after it, the better to spare the soldier's battledress which, by then, formed his 'best' uniform.

The question of a special uniform for ceremonial was considered soon after the end of the war. Re-introduction of full dress was financially impractical, other than for the Brigade of Guards, but something smarter than battledress was thought necessary for more important occasions. In 1947 such a uniform was introduced, based on the all-blue costume worn at the 1937 Coronation, and called No. 1 Dress. For most Line regiments this consisted of a jacket with a standing collar and collar badges, five-button fastening, pleated breast pockets and shoulder straps piped in the regimental facing colour, except for Royal regiments which had red. On the civilian-style trousers the old $\frac{1}{4}$ in. red welt of the full dress pattern was expanded to a 1 in. red stripe. As a headdress the coloured and peaked forage cap was reintroduced. Light Infantry had a dark green cap and jacket with silvered buttons, but blue trousers with a green stripe. Rifles had the whole uniform in their traditional colour of rifle green with black buttons. Lowland regiments had a dark blue doublet with Inverness skirts (rifle-green for the Cameronians), trews and spats. Highlanders had a differently cut doublet, in 'piper-green', which stopped short at the waist in front, but had short skirts behind, with turn-backs in the facing colour, the whole jacket rather in the cut of the Highland jacket worn before 1855. This, of course, was worn with kilt, hose, spats, and a small sporran of the eighteenth century type. Since both of the Scottish doublets showed six buttons above the waist and had no breast pockets, the whole ensemble, added to the tartan, had a more military and ceremonial air than the costume allotted to other regiments, which was not much different to the uniforms worn by various non-military services and organisations. As headdress, Scottish regiments had a dark blue bonnet with red tourie and dicing if applicable, or the Glengarry. The Cameronians' bonnet was rifle-green with a black tourie.

In times of economic stringency it is one thing to introduce a new uniform and quite another to issue it on a wide scale. All Regular officers had to purchase the uniform, sergeants, bandsmen and drummers (or their equivalent) were issued with it, but generally speaking it was only served out to the rank and file for very special parades, or for guards of honour, and withdrawn afterwards. The only occasion on which it was seen in large numbers was the Coronation of Queen Elizabeth II in 1953.

The waistbelt intended for No. 1 Dress was a coloured web girdle with circular clasp. However, this was unsuited to take a bayonet frog and thus the old Slade-Wallace type belt re-appeared. Officers wore either the Sam Browne or, in ceremonial order, a crimson waist sash under which was concealed a webbing belt supporting two

---

Plate 30: 1940–1962

88. Private, 1st Battalion, The Lancashire Fusiliers, 1944 (Burma). 89. Private, 1st Battalion, The Duke of Wellington's Regiment (West Riding), 1953 (Korea). 90. Rifleman, 2nd Battalion, 2nd King Edward VII's Own Goorkhas (Sirmoor Rifles), 1960 (Far East). Operations in the Far East, during the war and subsequently, saw the introduction of jungle- or olive-green clothing, as more suited to the terrain than khaki drill. For the Burma campaign, the slouch hat, worn at the turn of the century, was resurrected (Fig. 88). The 1937 equipment he wears was superseded in the Far East by the 1944 pattern in green webbing, with larger pouches and small pack and lighter water bottle (Figs. 89 and 90). Olive-green or battledress was worn during the early stages of the Korean War; later, 'combat clothing' in grey-green gaberdine material was issued. Headdress was the steel helmet, woollen 'cap comforter', or peaked cap with ear flaps in the same material (Fig. 89). This soldier is in patrol order with stripped down 1944 equipment of belt and water bottle. He carries a cosh, the later model of the Sten with its spare magazines in the jacket pockets, and wire-cutters in a 1937 webbing case on the belt. Short puttees and boots with moulded rubber soles are worn. The jungle warfare kit of olive-green hat with regimental identification signs, shirt or bush jacket, trousers and canvas jungle boots, worn during the Malaysian and Borneo campaigns, is shown on the Gurkha rifleman (Fig. 90). He has the 1944 equipment and 7.62 self-loading rifle.

Officer and corporal, 6th Battalion Durham Light Infantry in Normandy, 1944, wearing battledress. Both have 1937 equipment, the officer's with a basic pouch added for spare magazines for the Sten SMGs carried by both men. The battledress sleeves have regimental shoulder titles and formation signs.

gold-laced red leather slings for the sword. The sword was resumed for parade occasions in the early 1950s. Officers' shoulder straps were detachable so that, in ceremonial order, they could be replaced by gold shoulder cords of the type worn on the full dress tunic.

The 1937 pattern equipment continued in use throughout the war and for more than a decade after it. Various shades of *blanco* were applied to the webbing to keep it clean, including a dark greyish-green colour, a lighter khaki green and a yellowish-buff shade. During the war, particularly in the Middle East, the webbing was simply scrubbed until it achieved a near-white hue. In 1944 another equipment, in permanently dyed jungle-green webbing, was introduced for use in the Far East. This was roughly similar to the 1937 pattern but had larger pouches, small pack, and an aluminium water bottle with webbing carrier which clipped on to eyelets in the waistbelt. Two straps were attached to the underside of the small pack for securing the rolled-up poncho or groundsheet. Items for an officer's use, such as revolver holster, binocular case, compass and ammunition pouch, were issued with this equipment. The 1944 pattern did not become available for issue until World War 2 was over but it was used throughout the post-war Far East campaigns and in Kenya, and was also adopted for world-wide use by the Parachute Regiment. Together with the 1937 pattern, it remained in service until superseded by the next set of equipment, the 1958 pattern. As the latter is still in use today, it will be considered under the final period.

In 1939 the infantryman's personal weapon was still the short Lee-Enfield rifle but the requirements of mass production during the war, brought in another mark of it, a more easily manufactured weapon, known as the No. 4 Rifle, a simplified version of its predecessor. This, at first, was furnished with a very short socket-type spike bayonet. Later this blade reverted to a sword shape but of similar length. The No. 4 Rifle, and a lighter pattern for jungle operations, the No. 5, continued in service after the war until the general issue, towards the end of this period, of the first automatic rifle, the Belgian F.N., or self-loading rifle (SLR), of 7.62 mm calibre. Wartime conditions had demonstrated the need for a lighter more convenient weapon than a rifle, for platoon officers, NCOs, drivers, signallers and other specialists, hence the introduction of the sub machine-gun or carbine, of which the first, adopted in 1940, was the American Thompson (see Appendix 4).

After the war, the officers' chief uniform remained battledress, to which No. 1 Dress was added. Service dress and later mess dress were allowed on an optional basis. When the blue beret was adopted for the men, officers of most regiments resumed the pre-war service-dress cap in battledress. In some regiments, particularly Light Infantry and Rifles, the coloured field service cap made a re-appearance. All ranks of the Parachute Regiment wore the maroon beret in all orders of dress.

The greatest change in tropical uniform during World War 2, was the realisation, after nearly a hundred years, that the sun helmet, or topee, was unnecessary even under the hottest sun. It was, therefore finally discarded and home service headdress was used, except when steel helmets were required. The war in Burma witnessed a return of the slouch hat, or bush hat as it came to be called. This too was discarded after the war and, for operations and training in the Far East, a floppy hat of jungle-green drill, somewhat reminiscent of a small child's sun bonnet, was issued. Later an identical hat in khaki drill was introduced for the Middle East.

The pre-war khaki drill tunic was replaced by either an ordinary shirt or the so-called bush jacket, the latter resembling a shirt but with skirts which could be worn outside the trousers. Lower garments were either shorts, hose-tops and short puttees, or long trousers, worn loose, with anklets or short puttees. For service in the Far East a new colour of drill material was devised — jungle (later olive) green. A special kind of jungle boot made of canvas and rubber appeared after the war.

Once peace returned, tropical clothing underwent the usual tidying-up process, and for parades it came to achieve a smartness more reminiscent of pre-war days than was possible with battledress. When wearing shorts, many regiments took the opportunity of changing their khaki hose-tops for ones in a regimental colour with coloured garter flashes. Some Royal regiments, for example, had dark blue, while Light Infantry had dark green, the South Lancashires had maroon, the Essex had purple and the North Staffords and Northamptons had black, the latter with blue and buff garters. There were many other variations. Guards battalions had khaki hose-tops with crimson and blue bands round the turn-over.

When No. 1 Dress appeared at home, an all-white version in drill material was approved for ceremonial in hot climates — No. 3 Dress. Headdress and belts were as for No. 1 Dress. Owing to the comparative cheapness of such material on foreign stations, its use was more widespread than the use of No. 1 Dress on home stations.

Coronation contingent of the Northamptonshire Regiment in No. 1 Dress, 1953. The Colours are of the old pattern and were presented to the 58th Regiment in 1860. They were the last Colours in the Army to be carried in action, at the Battle of Laing's Nek in 1881.

The absorption of Gurkha regiments into the British Infantry brought with it two new types of headdress. In fact neither were truly new, for the broad and stiff-brimmed Gurkha hat with puggaree was merely a smarter version of the old slouch hat while the other cap was an updated Kilmarnock, of the type worn in undress by the British Infantry in the mid nineteenth century. The latter was rifle-green with a tourie and badge in front, the 2nd Goorkhas having a red and black diced band. Gurkha uniform was broadly the same as for British regiments, with Rifles appurtenances, their British officers having headdress similar to their counterparts in English Rifle regiments.

During World War 2 the Infantry wore battledress or the tropical clothing already described. In the North African and Italian campaigns khaki jerseys were often worn over KD shirts for extra warmth when the weather was not cold enough for battledress. The only completely original garment introduced in wartime was the Denison smock designed for the Parachute Regiment. Made of camouflaged material, it was pulled on over the head and covered the parts of the body that the abbreviated battledress blouse left exposed. To ensure a snug fit round the hips, the back skirt had a tail of material which passed between the legs to fasten with press studs on the inside of the front skirt.

In 1944 a new pattern steel helmet appeared as a replacement for the 1916 type which, with minor modifications to the lining and chin strap, had been in use until then. The new helmet sloped more steeply at the front than at the back and the brim was slightly curved at the sides, rather in the manner of a fifteenth century 'salade'. Parachute troops had a special basin-shaped helmet with no projecting brim and reinforced chin-strap.

During the Korean War the first British regiments to be committed wore jungle-green clothing in summer and battledress in winter. So harsh were the winters, however, that the latter had to be supplemented by fur-lined caps with ear flaps and other warm clothing, usually supplied by the United States Army. This hastened the need for a more satisfactory uniform than battledress and, in the later stages of the war, a new combat clothing

The Parachute Regiment, 1956, wearing maroon berets, Denison smocks and 1944 pattern equipment. The left-hand man carries a 3.5-inch rocket launcher.

was produced. Made of a windproof and water-repellent greyish-green gaberdine material, the jacket had numerous inner and outer pockets, zip and button fastenings and once again covered the upper body down to the hips with an all-round skirt furnished with draw-strings at waist and hip level. The trousers were of the same material as the jacket, with pockets similar to those on battledress. They were closed at the bottom either by anklets or, more popularly, by short puttees. A peaked cap with folding sides, similar to a ski cap, was issued with this clothing. For really cold weather, a 'parka' with synthetic fur lining and hood was additionally available. The boots, with heavily nailed leather soles, types of which had served the infantryman for generations, gave way to a new pattern with thick, moulded rubber soles. This clothing is shown in Fig. 89, Plate 30. Towards the end of this period, and the ending of National Service, this uniform began to be issued generally, to replace battledress for operations and training.

DRESS 1963–1980

Following the abolition of conscription and the resumption of voluntary enlistment, the opportunity was taken to re-clothe the new all-Regular Army in a more soldierly fashion, thus restoring the soldier's pride in his appearance, something that had been difficult to sustain in the battledress era. Essentially this was achieved by giving the soldier two basic uniforms; one for parade and one for operations and training. The latter initially took the form of the combat clothing mentioned in the preceding period. For the former, a soldier's new dress was based on the officer's service dress worn with collar and tie. The universal beret was relegated to being undress headgear, its place being taken for parade by the coloured and peaked forage cap designed for No.1 Dress, or its equivalent in Scottish and Irish regiments. Once the uniform for temperate climates had been settled, appropriate equivalents for tropical or warm weather use were designed.

In 1969 the first *Officers Dress Regulations* to be published since 1934 were issued, and these, with subsequent amendments, remain the basis of the Army's uniform today. The variations of the new uniform were codified into twelve different orders of dress, expanded from 1971 into fourteen. Although not all of these are applicable to soldiers (as opposed to officers), they will be used here to describe the uniform of the Infantry during this final period. This can only be done in general terms and it must be remembered that the traditional tendency of British regiments to modify, even ignore, Army regulations in the matter of dress still exists.

Corps of Drums, 2nd Battalion Royal Regiment of Fusiliers, 1974. Although fusilier caps and full dress tunics are worn, the latter lack the correct drummers' wings and lace.

No. 1 Dress — Temperate Ceremonial Uniform
This dress is as described in the preceding section, but from 1969 the regulations stated that 'it is not now in normal use for parade wear'. Other than to remark that it may still be seen worn by regimental bands on certain occasions, it will not be considered further.

No. 2 Dress — Service Dress (Temperate Parade and Ceremonial Uniform)
This, the modern soldier's 'best' dress, includes the coloured forage cap, skirted khaki jacket, shirt, collar and tie, khaki trousers, brown shoes for officers (except in Rifle regiments, which have black), and black boots for soldiers. Buttons for all are anodised, Rifles having black, Light Infantry silver, and the remainder gilt. (See Fig. 93, Plate 31.) Scottish regiments wear the Glengarry with No. 2 Dress, a jacket with cutaway skirts, trews or kilts with diced hose, their officers wear black shoes. The Royal Irish Rangers, from 1968 the only Irish Line regiment, have black buttons on their jackets, while their trousers and caubeens ('bonnets, Irish infantry') are in piper green. Fusiliers in No. 2 Dress wear the blue beret with regimental hackle. The Parachute Regiment wear the maroon beret. For less formal parades and daily duties, officers may wear the khaki service dress cap or, in some regiments, the coloured field service (or 'side') cap of the 1890 pattern. On ceremonial parades, officers are accoutred with Sam Browne belt and frog when swords are carried, soldiers with white waistbelt and bayonet frog. On the soldiers' parade belts the circular clasp of the type first introduced in 1850 may still be seen, but a rectangular clasp is also much in evidence. Rifles officers wear the black pouch belts and sling waistbelts formerly worn with full and No. 1 Dress, while their men retain their traditional black waistbelts. The Royal Irish Rangers are accoutred in a similar fashion to the Rifles.

7th Duke of Edinburgh's Own Gurkha Rifles in No. 6 Dress, warm weather parade uniform, 1976. From left: rifleman; company-sergeant-major; Gurkha officer; British officer. This clothing is described as 'stone-coloured'.

No. 3 Dress — Warm Weather Ceremonial Uniform
This is the white version of No. 1 Dress mentioned in the preceding section. In ceremonial order the white jacket is worn with the No. 1 Dress nether garments, but its use is now restricted in the same way as No. 1 Dress. The Foot Guards have a special No. 3 Dress in khaki drill.

No. 4 Dress — Warm Weather Uniform (Service Dress Pattern)
Only applicable to officers, this dress is the same as No. 2 Dress but in stone-coloured polyester and wool worsted.

No. 5 Dress
In the 1969 regulations this was listed as battledress, although this uniform was by then no longer worn by the Regular Army. It has now been deleted from the regulations.

No. 6 Dress — Warm Weather Parade Uniform (Bush Jacket Pattern)
This clothing, in stone-coloured polyester and cotton, is similar for officers and soldiers and equates to No. 2 Dress. The bush jacket has an open neck, with no collar and tie showing and the sleeves may be rolled above the elbows. It has anodised buttons and shoulder titles but no collar badges. Headdress and accoutrements are as prescribed for No. 2 Dress. Kilts or trews are permitted for Scottish regiments, whose jackets have cutaway skirts.

No. 7 Dress — Warm Weather Working Uniform
For daily routine in warm climates, a shirt, as used with No. 2 Dress, and lightweight olive-green trousers are ordered. Headdress for officers is the service dress cap, or equivalent; for soldiers, the beret or bonnet. Scottish infantry in this dress wear the 'bonnet TOS' (tam o' shanter) mentioned in earlier periods. It is customary to wear with this dress the web belt.

No. 8 Dress — Temperate Combat Uniform (Cold/Wet)
The clothing in this order of dress is the same for all ranks and is the successor to the combat

clothing first introduced at the time of the Korean War. This and No. 2 Dress form the soldier's two basic uniforms. From 1970 the greyish-green suit was replaced by a new camouflaged version in what is known as 'Disruptive Pattern Material (DPM)'. No. 8 Dress consists of a combat smock with liner and trousers and is completed by short puttees and 'boots, DMS' (direct moulded soles), as issued to soldiers. Recently, a higher boot, which enables the puttees to be dispensed with, has been tried out and will be generally issued in 1982–1983. Headdress for all ranks is either the steel helmet, a peaked DPM cap (with which a hood is also issued), the beret or bonnet.

The regulations state that if berets or bonnets are worn, regimental plumes or hackles, as used by Fusiliers, and by Scottish and Irish Infantry, should be discarded, but photographic evidence shows that this is frequently ignored. Berets are either dark blue, or rifle-green for the Light Infantry and Rifles, khaki for the Foot Guards and the Royal Anglian Regiment (the latter since 1976), or maroon for the Parachute Regiment.

No. 9 Dress — Combat Uniform (Jungle/Desert)

This is the equivalent of No. 8 Dress for hot climates and consists of a DPM tropical jacket and trousers, DPM hat, boots DMS with short puttees, or jungle boots.

No. 10 Dress — Temperate Mess Uniform

This and No. 11 Dress are only applicable to officers. The style and colour of the mess jacket accord to regimental pattern. The mess jacket is generally of the high-necked type, reminiscent of the old shell jacket, or it is of the roll collar type revealing shirt and bow tie. Jackets are scarlet with regimental facings which usually match the waistcoat. Regiments of the Light Division and Gurkha Rifles wear rifle-green. Whereas, before the war stiff collars and shirts were mandatory, soft shirts are now permitted except on ceremonial occasions or if civilians are present in tail coats and white ties. Scottish regiments wear their customary nether garments, but for the remainder either No. 1 Dress trousers or the 1939 pattern overalls with $\frac{1}{4}$ in. red welt are ordered.

No. 11 Dress — Warm Weather Mess Uniform

In this dress the temperate mess jacket is replaced by a white drill monkey jacket with either a waistcoat in the same material or a cummerbund of regimental pattern. Blue, and various shades of red or green are the most common colours for the cummerbund, but other varieties appear, such as primrose for the Gloucestershire Regiment, beech for the Staffordshire Regiment, red and rifle-green diced for the 2nd Goorkhas, while some Scottish regiments have theirs in tartan. Nether wear is as for No. 10 Dress.

No. 12 Dress — Coverall Clothing

This is simply a dress for dirty work when the uniform is covered by overalls.

No. 13 Dress — Barrack Dress

This is worn inside and outside barracks when the formality of No. 2 Dress is not required. It consists of a shirt with open neck worn under a 'jersey wool heavy', trousers of a dark greenish material, or kilt or trews for Scottish Infantry, and the appropriate headdress for the regiment concerned. Either a web belt or a coloured regimental-pattern stable belt is worn. The combat

---

Plate 31: 1963–1980

91. Sergeant, 1st Battalion, Royal Northumberland Fusiliers, 1967 (Aden). 92. Officer, The Argyll and Sutherland Highlanders (Princess Louise's) 1972. 93. Company Sergeant Major, The Royal Regiment of Wales, 1973. The abolition of conscription and return to an all-Regular Army saw the end of battledress and the 1937 equipment. The latter was replaced by the better balanced 1958 pattern, of which the belt, water bottle and one pouch are shown in Fig. 91. He is dressed for internal security duties in Aden. Fusiliers wear their regimental hackles in the coloured berets introduced in 1948. Shorts lost their popularity for tropical wear in favour of trousers and short puttees, worn here with a KD bush jacket tucked into the trousers. No.1 Dress, generally all blue, was authorised after the war as a smarter parade uniform than battledress but was never a general issue. Fig. 92 shows the Highland version, with short jacket in 'piper-green'. After battledress disappeared the soldier had two basic uniforms: combat clothing for operations and training (as in Fig. 89, Plate 30) and No. 2 Dress for other duties. Derived from the officer's service dress, the latter is worn with the coloured forage cap for ceremonial and parades (Fig. 93), or with the coloured beret.

91. Sgt, 1st Royal Northumberland Fus, 1967.

92. Offr, Argyll & Sutherland Hldrs, 1972.

93. C.S.M., Royal Regt of Wales, 1973.

94. Pte, Cheshire Regt, 1974.

96. L/Cpl, 1st Royal Anglian Regt, 1980.

95. Cpl-Bglr, Light Infantry, 1980.

smock may be added as extra protection against the weather.

No. 14 Dress — Shirt Sleeve Order

This is identical to No. 13 Dress but with the jersey omitted and shirt sleeves rolled to one inch above the elbow. Both are applicable to all ranks.

Guardsman, Grenadier Guards, in No. 8 Dress, temperate combat uniform (DPM), with 1958 pattern combat equipment fighting order and 7.62 mm self loading rifle, 1976.

These orders of dress, and the different varieties of clothing which constitute them, are applicable both to the Foot Guards and the Line Infantry, including the Gurkhas, but for the former there remains, in addition, full dress. This is authorised at public expense, for all ranks, for wear on Public Duties and certain ceremonial occasions. Its component parts, of bearskin cap, scarlet tunic and blue trousers are still essentially of the style and cut worn since 1856.

In recent years many Line regiments have acquired, at their own expense, full dress tunics for their bandsmen and drummers for wear on special occasions. In some instances these attempts to recapture something of the magnificence of the past are not always entirely satisfactory to the purist in these matters. The tunics sometimes lack the distinctive features laid down for bandsmen and drummers when full dress was last officially authorised before 1914, wings, drummers' lace, and so on. No. 1 Dress trousers frequently have to be substituted for the correct pattern, which had a more military cut and $\frac{1}{4}$ inch welts. While some regiments have acquired the proper blue helmets or fusilier caps, others have appeared in a variety of headgear of their own devising, such as converted police helmets or Wolseley helmets either in white or painted in a regimental colour. When the massed bands and drums of the Prince of Wales's Division beat Retreat on Horse Guards Parade in 1977, hardly two regiments' musicians were dressed alike, producing a variegated aspect among its nine

---

Plate 32: 1963–1980

94. Private, The Cheshire Regiment, 1974 (Northern Ireland). 95. Corporal-Bugler, The Light Infantry, 1980. 96. Lance-Corporal, 1st Battalion, The Royal Anglian Regiment, 1980. The Northern Ireland troubles have seen the addition of many items to the infantryman's equipment: batons, shields, riot guns, pocket radios, CS gas canisters, sniperscopes, plus the helmet visors and 'flak' jacket shown in Fig. 94. This soldier wears combat trousers and heavy wool jersey; the latter now forms part of the ordinary working dress in barracks. More recently the combat suit in DPM (disruptive pattern material) (Fig. 96) has been worn for all operations and training. A black plastic butt and stock has replaced the wooden parts of the SLR; one end of the sling is attached to the wrist to prevent the weapon being snatched. While working and fighting dress have increased in practicality, the traditional colours of scarlet, blue and green have re-appeared in bands, corps of drums, bugles and pipes, together, in some regiments, with blue helmets or fusilier caps. Fig. 95 is in The Light Infantry's No. 1 Dress with the Rifles pattern busby, adopted since 1973 by this regiment. Fig. 96 shows today's fighting dress and equipment, and General Purpose Machine Gun. The red and yellow strip on the arm brassard is a battalion distinguishing mark, the so-called 'Minden flash' of the former Suffolk Regiment. Although the blue beret remains the working headdress for most of the Infantry, the Royal Anglians have khaki, as do the Foot Guards; alternatively, a peaked DPM cap may be worn.

regiments, in marked contrast to the uniformity of similar assemblies at the pre-war Aldershot Tattoos, or the reviews before 1914, when only the badges and facing colours differentiated the helmeted regiments of the Line. The bands and bugles of the Royal Green Jackets have resurrected the 1890 pattern Rifles busby when wearing No. 1 Dress, a headgear also adopted by the sister regiment in the Light Division, the Light Infantry, and the Royal Irish Rangers.

Turning to equipment, the 1937 and 1944 patterns were superseded, as mentioned earlier, by the 1958 pattern, which provided two different assemblies: Combat Equipment Marching Order (CEMO) and Combat Equipment Fighting Order (CEFO), the former having a large pack. This equipment, like its predecessors, is made of webbing. It is dyed dark green, so that, like the 1944 pattern, it does not require blancoing, and its buckles and metal parts are painted green or black. The principle on which CEFO is based is that the load-carrying items are attached to a waistbelt resting on the hips, but supported over the shoulders by a broad padded yoke. The yoke meets between the shoulder blades from where two adjustable straps descend to be attached to the back of the waistbelt; at each of its two forward ends are quick-release buckles and two more straps, which pass through loops on the top of each ammunition pouch and return to the quick-release buckles. The ammunition pouches are clipped to the waistbelt forward of the hips, their tops level with the upper edge of the belt, unlike those of the 1937 and 1944 patterns which rode much higher. Attached to the rear sides of the left and right pouches are fittings to carry bayonet and rifle-grenade launcher respectively. Two 'kidney' pouches are fastened to the back of the belt by a quick-release device, these replace the small pack of the previous patterns. Below these hangs a cape carrier which clips to the lower edge of the belt and can also be clipped to the rearward bottom corners of the ammunition pouches. A plastic water bottle (with mug attached) in a web carrier goes on to the belt forward of one of the kidney pouches. In the equivalent vacant space on the other side the latest pattern of respirator can be hung. Finally, there are attachments at the rear juncture of the yoke and on the cape carrier for a lightweight pick or shovel. Now that the Infantry is mechanised, the large pack worn with CEMO is very rarely carried by a man, but if it has to be, it sits on the kidney pouches, attached by straps to loops on the yoke. A holster for the Browning 9 mm pistol, binocular and compass case are also available for this equipment. These comfortable and practical accoutrements, a great improvement on all preceding patterns and somewhat reminiscent of the 1871 Valise equipment in its weight-bearing principles, remain in service today. However, recent modifications and trials include enlargement of the original ammunition pouches, the substitution of a single load carrier for the kidney pouches, and the construction of the whole equipment in lightweight plastic material.

The steel helmet continues to be of the 1944 pattern, although new designs are under trial. For riot control in Northern Ireland a plastic visor has been attached to it, but for normal operations it is covered in camouflage material. Northern Ireland has also witnessed the return of body armour, not used since the seventeenth century, in the shape of the 'flak' jacket. The possibility of nuclear, biological or chemical warfare has necessitated the development of a NBC suit for wear over No. 8 Dress; the current pattern is the Mark III suit but an improved version in DPM is being developed.

Infantry weapons in use during this final period are listed at Appendix 4. Replacements for the SLR and GPMG in the 1980s are being developed and should provide two firearms very similar to one another in 4.85 mm calibre. Both the rifle, known as the Individual Weapon, and the machine gun, known as the Light Support Weapon, lack the traditional butts of their predecessors and are consequently much shorter, 30.3 inches and 35.4 inches respectively. These are not yet in service.

With the clothing and equipment developed during the last period of this survey, the British Infantry has once more regained the fine soldierly appearance it had always presented over the centuries, up to 1939. In addition, it now has the advantage, frequently absent in the past, of thoughtfully designed and well made combat uniforms and accoutrements. Perhaps the cut of the soldier's No. 2 Dress, particularly the trousers, leaves something to be desired, but measures are in hand to issue an improved pattern in the 1980s. There have been proposals that the improved pattern should be made in a dark green material, rather than khaki, but these have been set aside. Khaki, first devised by the British Army, is after all as much a 'national' military colour as scarlet once was, and since khaki has now been superseded for service by DPM, in the same way that khaki displaced scarlet at the turn of the century, it is logical and right that it should be preserved for parade wear.

Lieutenant-Colonel Luard, a noted dress reformer of the 1850s, wrote of scarlet: 'It has been worn in all our victories; it is known as the British colour all over the world. For troops of the line it is brilliant and imposing. It would be unwise to change it.' Developments in warfare which Luard could not have envisaged forced the change, though not until half a century later. Today only the Foot Guards can display to full advantage this once famous colour of the British Infantry. Yet even in an age of khaki or DPM, traces of it remain in the regiments of the Line; on the bands of forage caps of Royal regiments, on officers' mess jackets, on sergeants' sashes, as backing cloth for certain badges. Even in a Rifle regiment, one of the first to abandon it, traces of it can still be found. When the red-coated 60th Foot, raised in 1755, was converted to a Rifle corps early in the nineteenth century, its new green jacket retained scarlet as the facing colour. Today, as 2nd Battalion The Royal Green Jackets, its officers alone of the entire Infantry, are permitted scarlet backing to their rank insignia. Scarlet and rifle-green have passed into history, along with tricorne hats, grenadier caps, shakos and helmets. Facing colours and badges have altered with the reorganisations of the Infantry over the years. Accoutrements and weapons have developed in pace with the changing needs of the soldier's trade. It is far cry from the armoured pikeman of the seventeenth century to the mechanised infantryman of today, and yet, throughout the many changes the foot soldier's dress has undergone, there has always remained a thread of continuity and a distinctive character to it that is peculiarly British — martial without being militaristic. Much of this is the legacy of the regimental system, a treasured and envied system which has suffered many modifications but which continues to foster the infantryman's *esprit de corps*.

Men of 2nd Battalion Royal Green Jackets in NBC clothing (nuclear, biological, chemical) advancing from an Armoured Personnel Carrier 1980. The rifleman on the right is armed with the 84 mm Carl Gustav anti-tank weapon.

Appendix 1:

Regimental Facings and Lace Patterns of the Guards and Line According to the Regulations of 1742, 1751, 1768 and 1802

1. In the following lists, only those regiments which existed at the promulgation of the respective regulations, and which had an unbroken existence thereafter, are shown. Thus, the regiments numbered 50th–70th, raised from 1755 for the Seven Years War and which continued after 1763, do not appear until the 1768 column and their details prior to that date are not given, in fact, many are not known. The 1742 Clothing Book shows, in addition to those listed here, two regiments which took rank between the Invalids and the 43rd, later 42nd Highlanders, ten regiments of Marines and some independent companies. None of these are shown here, nor are those regiments raised after the 49th (later 48th Foot), but disbanded at the Peace of Aix-la-Chapelle in 1748. Regiments in the first column, shown with two numbers, i.e. (43rd) 42nd–(49th) 48th, assumed the second, permanent, number in 1748. Regiments numbered 71 and over, raised for the Seven Years War but disbanded in 1763 are not listed, nor are those raised for the American Revolution or the French Revolutionary and Napoleonic Wars, unless they survived the respective peaces with an unbroken existence. Thus, the regiments in the 1802 column numbered 71st and over, all of which survived throughout the nineteenth century, had had one or more previous existences. The number 95, for example, allotted in 1802 to the Rifle Corps which had been formed in 1800, had been borne by three former, quite separate regiments, from 1760–1763, 1780–1783 and from 1794–1796.

2. In 1742, the following regiments, the 1st–4th, 7th, 8th, 18th, 21st, 23rd and 27th, all had additional titles, e.g. 1st (Royal), 4th (King's Own), 18th (Royal Irish), 23rd (Royal Welch Fusiliers), etc. In 1782 most other regiments received subsidiary, territorial, titles, e.g. 6th (1st Warwickshire), 20th (East Devon), 54th (West Norfolk). Some of these titles changed subsequently, but many, although by no means all, were perpetuated in the 1881 territorial designations (see Appendix 3). To save space none of the additional titles are shown in this Appendix.

3. The details given under the columns headed *Men's lace* are the coloured patterns on the coat lace or braid. This was basically white, unless otherwise stated. The lace loops round the buttonholes in 1742 had either rounded, square or pointed ends; some were plain white. From 1751 all had square or pointed ends. To save space these distinctions are not given and reference must be made to C. C. P. Lawson's *History of Uniforms of the British Army* Vol. I, pp. 86–92 or the original sources. From 1768 all loops had square ends, except for a new style in bastion shape — regiments with this pattern are noted.

4. The colours given under the columns headed *Officers lace* also refer to their buttons. Prior to 1768 both gold and silver were used, but were not specified to regiments in the 1742 and 1751 regulations.

5. The heading *Buttons* in the 1802 column refers to the arrangement of lace loops and buttons on the coat fronts and cuffs after the abolition of lapels for the men in 1796. The same arrangement normally applied to buttons on officers' coats.

6. Abbreviations used in the Appendices:

blk	black
bl	blue
bu	buff
crim	crimson
fcgs	facings
Fus	Fusiliers
g	gold
gr	green
Gds	Guards
Hldrs	Highlanders
LI	Light Infantry
nc	no change
or	orange
prs	spaced in pairs
pur	purple
reg	spaced regularly
sca	scarlet
s	silver
threes	spaced in threes
wh	white
ye	yellow

Regt	*Facings 1742, 1751*	*Lace 1742*	*Lace 1751*
1st Gds	blue	plain	plain
Coldstream Gds	blue	plain	plain
3rd Gds	blue	plain	plain
1st Foot	blue	plain	plain
2nd	sea gr	bl/ye/bl bars	gr diagonals
3rd	bu	bu stripe, red chain	bl/red/bu stripes
4th	bl	bl zigzag	nc
5th[1]	gosling gr	plain	nc
6th	deep ye	red zigzag & arrows	nc
7th(Fus)	bl	red worm	bl stripe
8th	bl	2 bl stripes	ye stripe
9th	ye	red zigzag	bl/red/bl stripes
10th	bright ye	plain	2 bl zigzags, ye/red stripe
11th	full gr	ye/gr zigzag & arrows	2 red, 2 gr stripes
12th	ye	ye stripe	nc
13th[2]	philemot ye	ye zigzag	bl & red zigzag
14th	bu	(nil)	bl zigzag between red & bl stripes
15th	ye	(nil)	2 blk & ye stripes
16th	ye	red worm	ye zigzag between 2 red stripes
17th	grey/wh	2 bl zigzags	2 grey zigzags between 2 bl stripes
18th	bl	plain ye lace	bl scroll
19th	ye/gr	ye stripe, bl chain	2 red stripes between 2 bl stripes
20th	pale ye	plain	2 red stripes between 2 blk stripes
21st(Fus)	bl	bl zigzag, ye stripe	ye stripe, blk scroll
22nd	pale bu	bu stripe	2 stripes, red & bl dashes
23rd(Fus)	bl	ye & red indented stripes	ye & bl stripes, red diagonals
24th	willow gr	gr stripe	nc
25th	deep ye	red & bl zigzags	red & blk stripes
26th	pale ye	(nil)	2 ye stripes
27th	bu	2 ye worms	ye stripe between blk & bl zigzags
28th	bright ye	plain	blk diamond between 2 red stripes
29th	ye	(nil)	bl scroll between 2 bl & red stripes
30th	pale ye	plain	2 gr stripes
31st	bu	bu worm	gr stripe
32nd[2]	wh	wh worm, 2 blk stripes	blk zigzag between 2 blk stripes
33rd	red	plain	nc
34th	bright ye	bl leaves, ye dots	ye stripe, bl scroll
35th	or	plain	ye zigzag between ye & red stripes
36th	grass gr	gr coil	gr stripe
37th	ye	ye worm	2 ye stripes between red & bl zigzag
38th	ye	plain	ye stripe between 2 gr stripes
39th	gr	plain	gr scroll
40th	bu	ye & gr worms	2 bu/blk/bu stripes
(Invalids) 41st	bl	(nil)	(nil)
(43rd) 42nd (Hldrs)	bu	(nil)	2 red stripes
(44th) 43rd[2]	wh	bl stars	bl stars between red stripes
(45th) 44th	ye	(nil)	ye stripe between bl & blk zigzags
(46th) 45th	deep gr	gr stars	gr stripe & stars
(47th) 46th	ye	ye worm	2 blk zigzags, red stripe
(48th) 47th	wh	bl star & stripe	ye diamond between 2 blk zigzags
(49th) 48th	bu	ye stripe, gr chain	ye stripe between gr stripe & scroll
49th	full gr (1751)		ye stripe between 2 gr scrolls
50th			
51st[2]			
52nd[2]			
53rd			
54th			
55th			
56th			
57th			
58th			
59th			
60th			
61st			
62nd			
63rd			
64th			
65th			

Fcgs 1768	Offrs' lace 1768	Men's lace 1768	Fcgs 1802	Offrs' lace 1802	Men's lace 1802	Buttons 1802
nc	g	nc, bastion loops, reg, 1771	nc	nc	nc	reg
nc	g	nc, pointed loops, prs, 1773	nc	nc	nc	prs
nc	g	nc, pointed loops, threes, 1774	nc	nc	nc	threes
nc	g	double bl worm	nc	nc	nc	prs
bl	s	bl stripe	nc	nc	nc	reg
nc	s	ye/blk/red stripes	nc	nc	nc	prs
nc	s	bl stripe	nc	nc	nc, bastion	reg
nc	s	2 red stripes, bastion	nc	nc	nc	reg
nc	s	ye & red stripes	nc	nc	nc	prs
nc	g	nc	nc	nc	nc	reg
nc	g	bl & ye stripes	nc	nc	nc	reg
nc	s	2 blk stripes	nc	nc	nc	prs
nc	s	bl stripe	nc	nc	nc	reg
nc	g	2 red, 2 gr stripes, bastion	nc	nc	nc	prs
nc	g	ye/crim/blk stripes, bastion	nc	nc	nc	prs
nc	s	ye stripe	nc	nc	nc[3]	prs
nc	s	bl & red worm, bu stripe	nc	nc	nc[3]	prs
nc	s	ye & blk worm, red stripe	nc	nc	nc[3]	prs
nc	s	crim stripe	nc	nc	nc	reg
nc	s	2 bl/ye stripes	nc	nc	nc	prs
nc	g	bl stripe	nc	nc	nc	prs
deep gr	g	red & gr stripes	nc	nc	nc	prs
nc	s	red & blk stripes	nc	nc	nc	prs
nc	g	bl stripe	nc	nc	nc	prs
nc	g	bl & red stripes, bastion	nc	nc	nc	prs
nc	g	red/bl/ye stripes	nc	nc	nc, bastion	reg
nc	s	red & gr stripes	nc	nc	nc	prs
nc	g	bl/ye/red stripes, bastion	nc	nc	nc	reg
nc	s	1 bl, 2 ye stripes	nc	nc	nc	prs
nc	g	bl & red stripes	nc	nc	nc	reg
nc	s	1 ye, 2 blk stripes	nc	nc	nc	prs
nc	s	2 bl, 1 ye stripe, bastion	nc	nc	nc, square-end	prs
nc	s	sky-bl stripe, bastion	nc	nc	nc	reg
nc	s	bl & ye worm, red stripe	nc	nc	nc	reg
nc	g	blk worm & stripe	nc	nc	nc	prs
nc	s	red stripe, bastion	nc	nc	nc	prs
nc	s	bl & ye worm, red stripe	nc	nc	nc	prs
nc	s	ye stripe	nc	nc	nc	prs
gr	g	red & gr stripes	nc	nc	nc	prs
nc	s	red & ye stripes	nc	nc	nc	prs
nc	s	2 red, 1 ye stripe, bastion	nc	nc	nc[4]	reg
nc	g	light gr stripe	nc	nc	nc	prs
nc	g	red & blk stripe	nc	nc	nc	reg
nc	g	(nil)	red	s	blk stripe, bastion	reg
bl	g	red stripe, bastion	nc	nc	nc	reg
nc	s	red & blk stripe	nc	nc	nc	prs
nc	s	bl/ye/blk stripes	nc	nc	nc	reg
nc	s	gr stripe, bastion	nc	nc	nc	prs
nc	s	red & pur worms	nc	nc	nc	prs
nc	s	1 red, 2 blk stripes	nc	nc	nc	prs
nc	g	blk & red stripes	nc	nc	nc	prs
nc	g	2 red, 1 gr stripe, bastion	nc	nc	nc	reg
blk	s	red stripe	nc	nc	nc	prs
deep gr	g	gr worm, bastion	nc	nc	nc, square-end	prs
bu	s	red worm, or stripe	nc	nc	nc	prs
red	g	red stripe	nc	nc	nc	prs
popinjay gr	s	gr stripe	nc	nc	nc	prs
dark gr	g	2 gr stripes	nc	nc	nc	prs
pur	s	pink stripe	nc	nc	nc	prs
ye	g	blk stripe	nc	nc	nc	prs
blk	g	red stripe	nc	nc	nc	reg
pur	s	red & ye stripes	wh	g	2 blk stripes, bastion	reg
bl	s	2 bl stripes	nc	nc	nc	prs
bu	s	bl stripe	nc	nc	nc	reg
ye/bu	s	2 bl, 1 bu stripe	nc	nc	nc	prs
deep gr	s	gr stripe	nc	nc	nc	prs
blk	g	red & blk stripes	nc	nc	nc	prs
wh	s	red & blk worm, blk stripe	nc	g	nc	prs

Regt	*Facings 1742, 1751*	*Lace 1742*	*Lace 1751*
66th			
67th			
68th[2]			
69th			
70th			
71st (Hldrs)[2]			
72nd (Hldrs)[5]			
73rd (Hldrs)[5]			
74th (Hldrs)[5]			
75th (Hldrs)[5]			
76th			
77th			
78th (Hldrs)			
79th (Hldrs)			
80th			
81st			
82nd			
83rd			
84th			
85th[2]			
86th			
87th[1]			
88th			
89th			
90th[2]			
91st (Hldrs)[5]			
92nd (Hldrs)			
93rd (Hldrs)			
94th			
95th (Rifles)			
96th[6]			

Notes

[1] Converted to Fusiliers: 5th, 1836; 87th, 1827.

[2] Converted to Light Infantry: 13th, 1822; 32nd, 1858; 43rd and 52nd, 1803; 51st and 71st, 1809; 68th and 85th, 1808; 90th, 1815.

[3] Bastion lace by 1815.

[4] Square-end lace by 1815.

[5] Highland dress discontinued in 1809. 72nd resumed bonnet and trews in 1823, 74th and 91st resumed shako and trews in 1846 and 1864 respectively.

[6] Renumbered 95th when 95th Rifles became Rifle Brigade in 1816; disbanded 1818. 96th reraised in 1824.

Fcgs 1768	*Offrs' lace 1768*	*Men's lace 1768*	*Fcgs 1802*	*Offrs' lace 1802*	*Men's lace 1802*	*Buttons 1802*
ye/gr	g	crim & gr stripe, 1 gr stripe	nc	s	nc	reg
pale ye	s	ye, pur & gr stripes	nc	nc	nc	prs
deep gr	s	ye & blk stripes	nc	nc	nc	prs
willow gr	g	1 red, 2 gr stripes	nc	nc	nc	prs
blk	g	blk worm	nc	nc	nc	reg
			bu	s	red stripe	reg
			deep ye	s	gr stripe, bastion	reg
			dark gr	g	red stripe, bastion	reg
			wh	g	red & bl stripes	reg
			deep ye	s	2 ye, 1 red stripe	prs
			red	s	blk stripe	prs
			ye	s	blk stripe	reg
			bu	g	gr stripe, bastion	reg
			dark gr	g	1 ye, 2 red stripes	prs
			ye	g	2 red, 1 blk stripe	prs
			bu	s	bl & red stripes	prs
			pale ye	s	blk stripe, bastion	prs
			pale ye	g	gr & red stripes	prs
			pale ye	s	2 red stripes	prs
			ye	s	2 red worms, 2 blk stripes	prs
			ye	s	2 ye, 2 blk stripes	prs
			gr	g	red stripe	prs
			pale ye	s	2 blk, 2 red, 1 ye stripe	prs
			blk	g	red & bl stripes	prs
			deep bu	g	bl & bu stripes	prs
			ye	s	blk stripe, blk dart	prs
			ye	s	bl stripe	prs
			ye	s	ye stripe[3]	prs
			gr	g	red & gr stripes	prs
			blk	blk loops s buttons	nil (wh piping)	reg (3 rows)
			bu	s	red/ye/blk stripes	prs

Appendix 2:

Distinctions and Tartans for Guards, Fusiliers, Rifles and Scottish regiments, 1881–1914

Regiment	*Headdress*	*Tartan* (where appropriate)
Grenadier Guards	White plume, left of bearskin cap	
Coldstream Guards	Red plume, right	
Scots Guards	no plume	Royal Stewart (Pipers)
	Blue over red hackle (Pipers)	
Irish Guards	Blue, right	
Welsh Guards (1915)	White/green/white, left	
Royal Scots	Blackcock feathers[1]	Government[2] 1881–1901
		Hunting Stewart 1901
		Royal Stewart (Pipers)
Royal Northumberland Fusiliers	Scarlet over white, left[3]	
Royal Fusiliers	White, right[3]	
Lancashire Fusiliers	Primrose, left[3]	
Royal Scots Fusiliers	White, right[4]	Government[5]
Royal Welch Fusiliers	White, left[3/6]	
King's Own Scottish Borderers	Blackcock feathers[1]	Government[5] 1881–1897
		Leslie 1898
		Royal Stewart (Pipers)
Cameronians	Black[7]	Government[5] 1881–1891,
		Douglas 1892
Royal Inniskilling Fusiliers	Grey, left[3]	
Black Watch	Scarlet[8]	Government[5]
		Royal Stewart (Pipers)
King's Royal Rifle Corps	Scarlet over black[9]	
Highland Light Infantry	Green ball tuft[10]	Mackenzie (trews)
	Scarlet[8] (Bandsmen)	
Seaforth Highlanders	White[8]	Mackenzie
	Scarlet[8] (Bandsmen)	
Gordon Highlanders	White[8]	Gordon
Cameron Highlanders	White[8]	Cameron-Erracht
Royal Irish Rifles	Black and green[9]	
Royal Irish Fusiliers	Green, left[3]	
Argyll & Sutherland Highlanders	White[8]	Government[2]
Royal Munster Fusiliers	White and green, left[3]	
Royal Dublin Fusiliers	Blue and green, left[3]	
Rifle Brigade	Black[9]	

Notes

[1] From 1903 in Kilmarnock bonnet with diced band, scarlet-white-green.
[2] Or Sutherland.
[3] Plume in racoon-skin cap.
[4] Plume in sealskin cap.
[5] Or Black Watch.
[6] Black flash at back of collar.
[7] Plume in rifle-green shako, black cap-lines.
[8] Hackle in feather bonnet, diced band scarlet-green-white (Argylls, scarlet-white).
[9] Plume in black busby, black cap-lines.
[10] On dark blue (officers dark green, 1881–1900) shako with diced band, crimson-green-white, black cap-lines.

Appendix 3:

Numbered Regiments and Facings Prior to 1881, Their Subsequent Territorial Titles with Facings as at 1914, and their Present (1980) Titles

1. 94th — 100th Regiments
The 94th was revived in 1823 after being disbanded in 1818. The 95th — 99th were raised in 1824, the 100th in 1858.

2. 101st — 109th Regiments
All were taken into the Line in 1861 after the dissolution of the Honourable East India Company's armies following the Indian Mutiny (see notes[3] at end of this Appendix). The oldest, 103rd (Royal Bombay Fusiliers) was raised as Bombay Europeans in 1661, the 102nd (Royal Madras Fusiliers) was raised in 1748, and the 101st (Royal Bengal Fusiliers) in 1756. The remainder dated from between 1826 and 1854.

3. Rifle Brigade
The 95th Rifles were taken out of the Line in 1816 and styled the Rifle Brigade.

4. Territorial Titles
The numbered regiments were converted into two-battalion territorial regiments in 1881. Those numbered 1st–25th already had two battalions, so they merely exchanged their numbers for new titles; the remainder were grouped by pairs, except for the 79th, or Cameron Highlanders, which did not receive a second battalion until 1897, and the K.R.R.C. (60th) and Rifle Brigade which each had four battalions.

The regiments are listed under their most familiar names, but many had additional designations which underwent numerous permutations with the county titles; for example, in 1881 the 49th and 66th became the Princess Charlotte of Wales's (Berkshire Regiment) and the 57th and 77th because the Duke of Cambrige's Own (Middlesex Regiment), both reversing the bracketed portion in 1920.

5. Facings, 1914
In the fourth column are the facings worn in full dress in 1914, the last year in which it was officially worn by the whole Line. By this date several regiments had regained their old facing colours instead of the 'national' facings, of white (English and Welsh), yellow (Scottish) and green (Irish), ordered in 1881 for non-Royal regiments.

6. Titles 1980
The present-day titles in the last column reflect the reductions, amalgamations and formation of 'large' regiments carried out between 1958–1970. Only eleven, out of the seventy Line regiments in 1881, have survived this process.

From 1968 all regiments were grouped for administrative purposes into five 'Divisions'; these are also shown.

Besides the Line regiments in this column, the Regular Infantry in 1980 also includes:

Guards Division
Grenadier Guards (two bns)
Coldstream Guards (two bns)
Scots Guards (two bns)
Irish Guards (one bn)
Welsh Guards (one bn)

The Parachute Regiment (three bns) (formed 1942)

Brigade of Gurkhas
(From Indian Army 1948)
2nd King Edward VII's Own Goorkhas (Sirmoor Rifles) (two bns) (formed 1816)
6th Queen Elizabeth's Own Gurkha Rifles (one bn) (formed 1817)
7th Duke of Edinburgh's Own Gurkha Rifles (one bn) (formed 1902)
10th Princess Mary's Own Gurkha Rifles (one bn) (formed 1890).

All other existing regiments have one Regular battalion except; Queen's Regt, Royal Regt of Fusiliers, Royal Anglian Regt, Light Infantry, Royal Green Jackets (three each), Royal Irish Rangers (two).

Table follows overleaf

Former no.	*Fcgs*	*Territorial title*	*Fcgs in 1914*	*Present title and administrative division*
1st	bl	Royal Scots	bl	Royal Scots (Scottish)
2nd	bl	Queen's Royal Regt (West Surrey)	bl	Queen's Regt (Queen's)
3rd	bu	Buffs (Royal East Kent Regt) (1935[1])	bu	Queen's Regt (Queen's)
4th	bl	King's Own Royal Regt (Lancaster)	bl	King's Own Royal Border Regt (King's)
5th (Fus)	gosling gr	Royal Northumberland Fus (1935[1])	gosling gr	Royal Regt of Fusiliers (Queen's)
6th	bl (1832[1])	Royal Warwickshire Regt	bl	Royal Regt of Fusiliers (Queen's)
7th (Fus)	bl	Royal Fusiliers (City of London)	bl	Royal Regt of Fusiliers (Queen's)
8th	bl	King's Regt (Liverpool)	bl	King's Regt (King's)
9th	ye	Royal Norfolk Regt (1935[1])	ye	Royal Anglian Regt (Queen's)
10th	ye	Roya! Lincolnshire Regt (1946[1])	wh	Royal Anglian Regt (Queen's)
11th	Lincoln gr	Devonshire Regt	Lincoln gr	Devon & Dorset Regt (Prince of Wales's)
12th	ye	Suffolk Regt	ye	Royal Anglian Regt (Queen's)
13th (Ll)	bl (1842[1])	Somerset Ll	bl	Light Infantry (Light)
14th	bu	West Yorkshire Regt	bu	Prince of Wales's Own Regt of Yorkshire (King's)
15th	ye	East Yorkshire Regt	wh	Prince of Wales's Own Regt of Yorkshire (King's)
16th	ye	Bedfordshire & Hertfordshire Regt[2]	wh	Royal Anglian Regt (Queen's)
17th	wh	Royal Leicestershire Regt (1946[1])	wh	Royal Anglian Regt (Queen's)
18th	bl	Royal Irish Regt	bl	disbanded 1922
19th	grass gr	Green Howards (Yorkshire Regt)	grass gr	Green Howards (Yorkshire Regt) (King's)
20th	ye	Lancashire Fusiliers	wh	Royal Regt of Fusiliers (Queen's)
21st (Fus)	bl	Royal Scots Fusiliers	bl	Royal Highland Fusiliers (Scottish)
22nd	bu	Cheshire Regt	bu	Cheshire Regt (Prince of Wales's)
23rd (Fus)	bl	Royal Welch Fusiliers	bl	Royal Welch Fusiliers (Prince of Wales's)
24th	grass gr	South Wales Borderers	grass gr	Royal Regt of Wales (Prince of Wales's)
25th	bl (1805[1])	King's Own Scottish Borderers	bl	King's Own Scottish Borderers (Scottish)
26th	ye	1st Bn Cameronians (Scottish Rifles)	gr	disbanded 1968
27th	bu	1st Royal Inniskilling Fus (1881[1])	bl	Royal Irish Rangers (King's)
28th	ye	1st Gloucestershire Regt	wh	Gloucestershire Regt (Prince of Wales's)
29th	ye	1st Worcestershire Regt	wh	Worcestershire & Sherwood Foresters Regt (Prince of Wales's)
30th	ye	1st East Lancashire Regt	wh	Queen's Lancashire Regt (King's)
31st	bu	1st East Surrey Regt	wh	Queen's Regt (Queen's)
32nd (Ll)	wh	1st Duke of Cornwall's Ll	wh	Light Infantry (Light)
33rd	sca	1st Duke of Wellington's Regt (W Riding)	sca	Duke of Wellington's Regt (W Riding) (King's)
34th	ye	1st Border Regt	ye	King's Own Royal Border Regt (King's)
35th	bl (1832[1])	1st Royal Sussex Regt	bl	Queen's Regt (Queen's)
36th	grass gr	2nd Worcestershire Regt	wh	Worcestershire & Sherwood Foresters Regt (Prince of Wales's)
37th	ye	1st Royal Hampshire Regt (1946[1])	ye	Royal Hampshire Regt (Prince of Wales's)
38th	ye	1st South Staffordshire Regt	wh	Staffordshire Regt (Prince of Wales's)
39th	grass gr	1st Dorsetshire Regt	grass gr	Devon & Dorset Regt (Prince of Wales's)
40th	bu	1st South Lancashire Regt	wh	Queen's Lancashire Regt (King's)
41st	wh	1st Welch Regt	wh	Royal Regt of Wales (Prince of Wales's)
42nd (Hldrs)	bl	1st Black Watch (Royal Highland Regt)	bl	Black Watch (Royal Highland Regt) (Scottish)
43rd (Ll)	wh	1st Oxfordshire & Buckinghamshire Ll[2]	wh	Royal Green Jackets (Light)
44th	ye	1st Essex Regt	wh	Royal Anglian Regt (Queen's)
45th	Lincoln gr	1st Sherwood Foresters (Notts & Derby)	Lincoln gr	Worcestershire & Sherwood Foresters Regt (Prince of Wales's)
46th	ye	2nd Duke of Cornwall's Ll	wh	Light Infantry (Light)
47th	wh	1st Loyal Regt (North Lancashire)	wh	Queen's Lancashire Regt (King's)
48th	bu	1st Northamptonshire Regt	wh	Royal Anglian Regt (Queen's)
49th	Lincoln gr	1st Royal Berkshire Regt (1885[1])	bl	Duke of Edinburgh's Royal Regt (Prince of Wales's)
50th	bl (1831[1])	1st Royal West Kent Regt	bl	Queen's Regt (Queen's)
51st (Ll)	bl (1821[1])	1st King's Own Yorkshire Ll	bl	Light Infantry (Light)
52nd (Ll)	bu	2nd Oxfordshire & Buckinghamshire Ll[2]	wh	Royal Green Jackets (Light)
53rd	sca	1st King's Shropshire Ll	bl	Light Infantry (Light)
54th	grass gr	2nd Dorsetshire Regt	grass gr	Devon & Dorset Regt (Prince of Wales's)
55th	Lincoln gr	2nd Border Regt	ye	King's Own Royal Border Regt (King's)
56th	pur	2nd Essex Regt	wh	Royal Anglian Regt (Queen's)
57th	ye	1st Middlesex Regt	lemon ye	Queen's Regt (Queen's)
58th	blk	2nd Northamptonshire Regt	wh	Royal Anglian Regt (Queen's)
59th	wh	2nd East Lancashire Regt	wh	Queen's Lancashire Regt (King's)

60th (Rifles)	sca	King's Royal Rifle Corps	sca	Royal Green Jackets (Light)
61st	bu	2nd Gloucestershire Regt	wh	Gloucestershire Regt (Prince of Wales's)
62nd	bu	1st Wiltshire Regt	bu	Duke of Edinburgh's Royal Regt (Prince of Wales's)
63rd	Lincoln gr	1st Manchester Regt	wh	King's Regt (King's)
64th	blk	1st North Staffordshire Regt	wh	Staffordshire Regt (Prince of Wales's)
65th	wh	1st York & Lancaster Regt	wh	disbanded 1968
66th	grass gr	2nd Royal Berkshire (1885[1])	bl	Duke of Edinburgh's Royal Regt (Prince of Wales's)
67th	ye	2nd Royal Hampshire Regt (1946[1])	ye	Royal Hampshire Regt (Prince of Wales's)
68th (LI)	dark gr	1st Durham LI	dark gr	Light Infantry (Light)
69th	Lincoln gr	2nd Welch Regt	wh	Royal Regt of Wales (Prince of Wales's)
70th	blk	2nd East Surrey Regt	wh	Queen's Regt (Queen's)
71st (Hld LI)	bu	1st Highland LI	bu	Royal Highland Fus (Scottish)
72nd (Hldrs)	ye	1st Seaforth Hldrs	bu	Queen's Own Hldrs (Scottish)
73rd	dark gr	2nd Black Watch (Royal Highland Regt)	bl	Black Watch (Royal Highland Regt) (Scottish)
74th (Hldrs)	wh	2nd Highland LI	bu	Royal Highland Fus (Scottish)
75th	ye	1st Gordon Hldrs	ye	Gordon Highlanders (Scottish)
76th	sca	2nd Duke of Wellington's Regt (W Riding)	sca	Duke of Wellington's Regt (W Riding) (King's)
77th	ye	2nd Middlesex Regt	lemon ye	Queen's Regt (Queen's)
78th (Hldrs)	bu	2nd Seaforth Hldrs	bu	Queen's Own Hldrs (Scottish)
79th (Hldrs)	bl (1873[1])	Queen's Own Cameron Hldrs	bl	Queen's Own Hldrs (Scottish)
80th	ye	2nd South Staffordshire Regt	wh	Staffordshire Regt (Prince of Wales's)
81st	bu	2nd Loyal Regt (North Lancashire)	wh	Queen's Lancashire Regt (King's)
82nd	ye	2nd South Lancashire Regt	wh	Queen's Lancashire Regt (King's)
83rd	ye	1st Royal (Irish) Ulster Rifles	gr	Royal Irish Rangers (King's)
84th	ye	2nd York & Lancaster Regt	wh	disbanded 1968
85th (LI)	bl (1821[1])	2nd King's Shropshire LI	bl	Light Infantry (Light)
86th	bl (1812[1])	2nd Royal (Irish) Ulster Rifles	gr	Royal Irish Rangers (King's)
87th (Fus)	bl (1827[1])	1st Royal Irish Fus	bl	Royal Irish Rangers (King's)
88th	ye	1st Connaught Rangers	gr	disbanded 1922
89th	blk	2nd Royal Irish Fus	bl	Royal Irish Rangers (King's)
90th (LI)	bu	2nd Cameronians (Scottish Rifles)	gr	disbanded 1968
91st (Hldrs)	ye	1st Argyll & Sutherland Hldrs	ye	Argyll & Sutherland Hldrs (Scottish)
92nd (Hldrs)	ye	2nd Gordon Hldrs	ye	Gordon Highlanders (Scottish)
93rd (Hldrs)	ye	2nd Argyll & Sutherland Hldrs	ye	Argyll & Sutherland Hldrs (Scottish)
94th	Lincoln gr	2nd Connaught Rangers	gr	disbanded 1922
95th	ye	2nd Sherwood Foresters (Notts & Derby)	Lincoln gr	Worcestershire & Sherwood Foresters Regt (Prince of Wales's)
96th	ye	2nd Manchester Regt	wh	King's Regt (King's)
97th	sky bl	2nd Royal West Kent Regt	bl	Queen's Regt (Queen's)
98th	wh	2nd North Staffordshire Regt	wh	Staffordshire Regt (Prince of Wales's)
99th	ye	2nd Wiltshire Regt	bu	Duke of Edinburgh's Royal Regt (Prince of Wales's)
100th	bl	1st Prince of Wales's Leinster Regt	bl	disbanded 1922
101st (Fus)[3]	bl	1st Royal Munster Fus	bl	disbanded 1922
102nd (Fus)[3]	bl	1st Royal Dublin Fus	bl	disbanded 1922
103rd (Fus)[3]	bl	2nd Royal Dublin Fus	bl	disbanded 1922
104th (Fus)[3]	bl	2nd Royal Munster Fus	bl	disbanded 1922
105th (LI)[3]	bu	2nd King's Own Yorkshire LI	bl	Light Infantry (Light)
106th (LI)[3]	wh	2nd Durham LI	dark gr	Light Infantry (Light)
107th[3]	wh	2nd Royal Sussex Regt	bl	Queen's Regt (Queen's)
108th[3]	pale ye	2nd Royal Inniskilling Fus	bl	Royal Irish Rangers (King's)
109th[3]	wh	2nd Prince of Wales's Leinster Regt	bl	disbanded 1922
Rifle Brigade	blk	Rifle Brigade	blk	Royal Green Jackets (Light)

Notes

[1] Granted Royal title/blue facings in year shown.

[2] 'Hertfordshire' added 1919; 'Buckinghamshire' 1908.

[3] Prior to 1861 these were European regiments of the Honourable East India Company's Armies.
101st and 104th = 1st and 2nd European Bengal Fusiliers; 107th = 3rd European Bengal LI; 102nd = 1st Madras (European) Fusiliers; 105th = 2nd Madras (European) LI; 108th = 3rd Madras Europeans; 103rd = 1st Bombay (European) Fusiliers; 106th = 2nd Bombay (European) LI; 109th = 3rd Bombay Europeans.

Appendix 4:

Infantry Weapons 1660–1980

MUSKETS, RIFLES AND BAYONETS

The chief characteristics of the principal small arms used by the Regular Infantry are listed in the table following these notes.

SWORDS

Officers

The regulation pattern swords carried by Infantry officers, between the abolition of the spontoon as their principal weapon in 1786 and the present day, have been those of: 1786, 1796, 1803, 1822 (new blade in 1845), 1892 and 1895. The latter is still used today for ceremonial purposes, having last been carried in action in 1914. Guards, Flank Companies, Rifles and Scottish regimental officers have, at certain periods, had swords differing from the universal infantry pattern.

Soldiers

Swords were carried by all ranks of Infantry until 1768, by all sergeants and Grenadier companies until 1784, by all sergeants until 1852, thereafter only by staff-sergeants until 1914. From 1796 the regulation sergeants' sword resembled the officers' current pattern, until a special staff-sergeants' sword was authorised in 1905, modified in 1912. Guards, Rifles and Scottish regiments had their own patterns. Swords of various patterns have been carried by drummers, buglers, pipers, bandsmen and pioneers.

AUTOMATIC WEAPONS

Machine Guns

Two Maxim machine guns were issued to each battalion from 1890. The final version of the Maxim, known as the Vickers-Maxim, later the Vickers medium machine gun, appeared in 1912 and remained in service until 1963.

Approximate dates	*Type*	*Weight*	*Barrel length (inches)*	*Calibre*	*Rate of fire*	*Rounds carried*
1660–1700	Matchlock	13½ lb	42	14-bore	1 in 2 min	25
	Flintlock	–	42–46	14-bore	1–2 per min	25
FLINTLOCKS, MUZZLE-LOADING						
1725	Long land	10 lb 12 oz	46	.75 in.	3 per min	24/36
1769	Short land	10 lb	42	.75 in.	3 per min	56 (1784)
1794	India patt	9 lb 1 oz	39	.75 in.	3 per min	60 (1808)
1803–1839	New land service	10 lb 4½ oz 10 lb 1½ oz	42 39	.753 in.	3 per min	60
1800	Baker rifle	9 lb ½ oz	30	.625 in.	1 per min	80
PERCUSSION, MUZZLE-LOADING						
1839	Flintlock conversion	10 lb	39	.753 in.	2 per min	60
1842	Pattern	10 lb 3 oz	39	.753 in.	2 per min	60
1837	Brunswick rifle	9 lb 6 oz	30	.704 in.	2 per min	60
1851	Minie rifle	10 lb	39	.702 in.	2 per min	60
1853	Enfield rifle	8 lb 14½ oz	39	.577 in.	2 per min	60
	Enfield rifle (short)	8 lb 4½ oz	33	.577 in.	2 per min	60
METAL CARTRIDGES, INTERNAL IGNITION, BREECH-LOADING						
1866	Snider rifle	9 lb 5 oz	39	.577 in.	10 per min	70
	Snider rifle (short)	8 lb 11 oz	33	.577 in.	10 per min	70
1874	Martini-Henry rifle	8 lb 10½ oz	33	.45 in.	12 per min	70
MAGAZINE RIFLES, SMOKELESS POWDER						
1888–1892	Lee-Metford, Mks I & II	9 lb 8 oz	30	.303 in.	12 per min; 1 in 2½ secs using magazine	100
1895	Lee-Enfield	9 lb 4 oz	30	.303 in.	12 per min; 1 in 2½ secs using magazine	100
1902–1914	Short Lee-Enfield, various marks	8 lb 10½ oz 8 lb 2½ oz	25.19	.303 in.	15 per min, normal; 20–25 per min, rapid	150–250
1940–1946	Lee-Enfield No. 4	9 lb 1 oz	25.19	.303 in.	15 per min, normal; 20–25 per min, rapid	100–250
SELF-LOADING RIFLE						
1954	F.N. S.L.R.	9.2 lb	21	7.62 mm	30–40 per min	60 in 3 magazines

Light Machine Guns
The Lewis Gun was introduced in 1915. It was superseded by the Bren light machine gun from 1937 which remained in service until 1963.
General Purpose Machine Gun
The G.P.M.G. replaced both the M.M.G. and the L.M.G. from 1963, and remains in service today.
Sub Machine Guns/Carbines
From 1940 the American Thompson sub machine gun was issued to junior commanders in platoons to replace revolvers and rifles. At the same time, the British Sten was developed and remained in service until 1953, when it was superseded by the Sterling, still in use today.

MORTARS

Battalion Weapons
The Trench howitzer, Marks I and II, was introduced in 1914. The Stokes mortar was in service from 1916–1919 but was withdrawn from battalions to brigade companies.
The 3-inch mortar was adopted as a battalion weapon from 1936. It was replaced in 1962 by the 81 mm mortar, still in service today.
Platoon Weapons
The 2-inch mortar was introduced during World War 2 and is due to be replaced by a 51 mm model.

ANTI-TANK WEAPONS

Battalion Weapons
1944–1949 6-pounder gun
1948–1957 17-pounder gun
1952–1956 B.A.T. (Battalion Anti-Tank weapon, recoil-less)
1957–1962 MOBAT
1962–1970 WOMBAT
1970– CONBAT
Platoon Weapons
1936–1945 Boys anti-tank rifle
1942–1953 Projector infantry anti-tank (PIAT)
1953–1958 3.5-in. rocket launcher
1958– 84 mm (Carl Gustav)
1970– 66 mm

Length, type of bayonet	*Sighted to (yds)*	*Remarks*
12 in. plug (1672)	–	Black-painted stock
Plug & ring (1690)	–	Paper cartridges with bullet & powder from 1686
17 in. socket	–	'Brown Bess'; brown polished stock; steel ramrod from 1724
17 in. socket	–	
17 in. socket	–	War economy pattern; 6 ft long with bayonet
17 in. socket	–	39 in. browned barrel with backsight for LI from 1810; all barrels browned 1820
23 in. sword	200	Issued to 95th Rifles and 5th Bn 60th Foot
17 in. socket	–	75 caps; backsight
17 in. socket	–	75 caps; backsight
22 in. sword	300	Rifle regiments; belted ball round
17 in. socket	1,000	Issued to all Infantry; conical bullet
17 in. socket	1,200	75 caps; effective range 250 yd
$22\frac{3}{4}$ in. sword	1,000	75 caps, issued to Sgts and Rifle Regts
17 in. socket	1,000	Single-loader; brass cartridge
$22\frac{3}{4}$ in. sword	1,000	Issued to Sgts and Rifle Regts
22 in., both types	1,450	Single-loader; last using black powder; sword bayonets for Sgts and Rifle Regts
12 in. sword	2,800	Cut-off to magazine for single-loading; magazine 8 rounds (1888); 10 rounds (1892); bolt-operated
12 in. sword	2,800	10 rounds in magazine
17 in. sword	2,000	10-round magazine; loaded by 5-round charger
8 in. spike	1,300	Wartime model; later with short sword bayonet
short sword	600	20-round magazine; wooden parts replaced by black plastic from 1975

Appendix 5:

Colours

Although Colours cannot be regarded as uniform, they have always been a central feature of an infantry regiment's overall appearance. Carried by ensigns and escorted by sergeants, they provided a focus and rallying point for a regiment on the battlefield and parade ground. The last Colours to be carried in action were those of the 58th Regiment at the Battle of Laing's Nek in 1881 but they still occupy an honoured position in a battalion's ceremonial today. The only exceptions are Rifle regiments which have never had Colours. For many years Colours were issued to regiments as 'Accoutrements', so a brief word on their general design is given here.

SEVENTEENTH CENTURY

From the time of the Civil War it was common practice for each company of an English regiment of Foot, including those of which the colonel, lieutenant-colonel and frequently the major were additionally their captains, to have a Colour. The colonel's was of a pure plain hue; the lieutenant-colonel's had the same ground as the colonel's but with the St George's Cross in the upper corner next to the pike; the major's was the same but with a pile wavy issuing from the upper corner; all captains' also with the Cross but each with varied devices. Scottish regiments had the St Andrew's Cross over the whole ground of the Colour.

At the Restoration this system continued, except that the St George's Cross came to cover the whole Colour except for the colonel's. In the 1st Guards the senior company Colour was the King's, which in due course was plain crimson with the Royal cypher and crown in gold. Charles II also presented Royal badges for each company of the 1st Guards, such badges not being granted to the Coldstream and 3rd Guards until 1750. The dimensions of the 1st Guards Colours in 1684 were 6 ft 9 in. flying and 6 ft 6 in. on the pike.

The basic ground colour for a Foot regiment's Colours was a matter for its colonel.

Under William III the number of Colours in a regiment was reduced to those of the three field officers, except in the Foot Guards. During the next reign it was, generally, further reduced to the colonel's and the lieutenant-colonel's.

EIGHTEENTH CENTURY

On the Act of Union in 1707 Scottish regiments lost their own national Colours and the St Andrew's Cross was joined to the St George's Cross on all Colours to form the Union. Soon afterwards, colonels' Colours acquired a small Union in the upper corner, leaving the field of the Colour to display their arms or some special badge. Lieutenant-colonels', majors' and, in the Guards, captains' Colours continued to have the Union throughout.

With the issue of the 1747 regulations the system of field officers' Colours was abolished for regiments of Foot, in favour of an entirely new arrangement of two Colours per regiment or battalion without any personal devices of its colonel permitted.

The size was to be 6 ft 6 in. flying by 6 ft 2 in. on the pike, which itself was 9 ft 10 in. long, including the 4 inch spear head, from which hung two 3 ft crimson and gold cords with tassels. The First, or King's Colour, was the Union flag, the Second, later called the Regimental Colour, was in the facing colour, with the Union in the upper canton, except for regiments faced red or white, which were to have additionally the St George's Cross with the Union in the upper canton. In the centre of both Colours was the number of the regiment in Roman characters, within a rose and thistle wreath, except for Royal regiments and the Six Old Corps, which had their special devices in the centre and their numbers in the upper corner.

Guards regiments also had only two Colours per battalion, but their design adhered to the old system, thus making Guards' Colours quite different to the Line's. They have remained different to the present day. For the Guards the former field officers' Colours became the King's Colours of each battalion in seniority, the two junior Colours having the Union returned to the upper corner, the ground, like the former colonels', being crimson. In the centre the following were emblazoned

1st Guards: The crown (1st Bn); the crown and Royal cypher, reversed and interlaced (2nd and 3rd Bns)

Coldstream Guards: The crown, garter and star (1st Bn); the crown and star within the garter (2nd Bn); the crown and garter star (3rd Bn).
3rd Guards: the crown, with the arms of Scotland and the motto *En Ferus Hostis* (1st Bn); the crown, the rose and thistle conjoined (2nd Bn); the crown and the star of the Order of the Thistle (3rd Bn).

For the Regimental Colours of each Guards battalion, one of the former captains' company Colours (the Union throughout with Royal badge) was borne in rotation.

The King's company Colour of the 1st Guards (applicable only to that regiment) was the Royal standard of the regiment; plain crimson, the Royal cypher reversed and interlaced surmounted by the crown, the rose, thistle, fleur-de-lys and harp, all crowned in each corner.

The 1768 warrant gives the same dimensions for Colours except that they were to be 6 ft on the pike. The Regimental Colour of regiments faced black was to have the St George's Cross throughout, the Union in the upper canton, the other three black. The regimental number in the centre of each Colour was placed on a cartouche, later replaced by a red shield.

The first battle honour to be awarded, GIBRALTAR, was ordered, in 1784, to be placed beneath the number on the Regimental Colours of the 12th, 39th, 56th and 58th Regiments.

NINETEENTH CENTURY

In 1801 the Act of Union with Ireland added the Cross of St Patrick to the Union, making it the design in use today, and the shamrock was entwined with the rose and thistle to form the Union wreath.

Many battle honours were awarded after the Napoleonic Wars and from 1820 these were borne on both Colours. The central shield gave way to a circular girdle, inscribed with the territorial titles awarded in 1782, the number in Arabic characters within.

From 1839 Roman numerals were reintroduced to replace Arabic in the centre of each Colour.

In 1844 the Queen's Colour was to bear no other device than the number in gold surmounted by the crown in the centre of the Union. Battle honours were therefore to be borne only on the Regimental Colour.

In 1855 the dimensions were reduced to 6 ft flying, 5 ft 6 in. on the pike. In 1858 they were further reduced to 4 ft 6 in. by 4 ft.

In 1868 the size was again reduced to 3 ft 9 in. flying, 3 ft deep. Each Colour was to be ornamented with fringe (gold and silver for the Queen's, gold and white for the Regimental). The Royal crest replaced the spearhead on the pikes.

With the reorganisation of the Line into territorial regiments in 1881, the honours awarded to the former regiments were henceforth to be borne on the Regimental Colours of both battalions, arranged on a circular laurel wreath. The small Union in the upper canton of the Regimental Colour was removed. In the centre of the Regimental Colour, within the Union wreath and below the crown, was a girdle inscribed with the new designation and encircling either the special badge (as borne, for example, by those formerly known as the Six Old Corps), or any additional part of the designation, or the number of the battalion in Roman figures. The alternative position for the latter was in the upper corner. Any other devices to which a regiment was entitled went in the other corners. Regiments faced white had the St George's Cross throughout.

TWENTIETH CENTURY

In 1919 each regiment was permitted to bear ten of the honours awarded for World War I on the King's Colour. This also applied to Guards' Colours which had continued unchanged except for the reduction in size and the addition of honours to their Regimental Colours.

In 1930 any regiment not already authorised to display a special badge in the centre of the Regimental Colour was permitted to do so.

After World War 2 a further ten honours were added to the King's Colour.

On the amalgamation of regiments, or formation of 'large' regiments from 1958, all former honours accrued to the new regiment. Only honours for the two World Wars were permitted on the Queen's Colour.

Bibliography

Anderson, D.N. and McKay, J.B., *The Highland Light Infantry, 1881–1914*, privately published, 1977.
Barnes, Major R.M., *A History of the Regiments and Uniforms of the British Army*, Seeley Service, London, 1950.
Barnes, Major R.M., *Uniforms and History of the Scottish Regiments*, Seeley Service, London, 1960.
Barnes, Major R.M., *Military Uniforms of Britain and the Empire*, Seeley Service, London, 1960.
Barnes, Major R.M., *The Soldiers of London*, Seeley Service, London, 1963.
Barthorp, Michael, *The Armies of Britain*, 1485–1980, National Army Museum, London, 1980.
Barthorp, Michael, *Crimean Uniforms—British Infantry*, Historical Research Unit, London, 1974.
Bowling, A.H., *British Infantry Regiments, 1660–1914*, Almark, New Malden, 1970.
Bowling, A.H., *Scottish Regiments, 1660–1914*, Almark, New Malden, 1970.
Bowling, A.H., *The Foot Guards Regiments, 1880–1914*, Almark, New Malden, 1972.
Blackmore, Howard L., *British Military Firearms, 1650–1850*, Herbert Jenkins, London, 1961.
Carman, W.Y., *British Military Uniforms*, Leonard Hill, Glasgow, 1957.
Carman, W.Y., *Costume of the 46th Regiment, 1837, by M.A. Hayes*, National Army Museum, London, 1972.
Carman, W.Y., *A Dictionary of Military Uniform*, Batsford, London, 1977.
Chandler, D.G., *The Art of War in the Age of Marlborough*, Batsford, London, 1976.
Chappell, M., *British Infantry Equipments, 1808–1908 and 1908–1980*, Osprey, London, 1980.
Chichester, H.M. and Burges-Short, G., *The Records and Badges of every Regiment and Corps in the British Army*, William Clowes, London, 1895.
Cooper-King, Colonel C., *The British Army and Auxiliary Forces*, 2 Vols, Cassell, London, 1893.
Featherstone, Donald, *Weapons and Equipment of the Victorian Soldier*, Blandford Press, Poole, 1978.
Forbes, Major-General A., *A History of the Army Ordnance Services*, 3 Vols, Medici Society, London, 1929.
Fortescue, Sir John, *History of the British Army*, 13 Vols, Macmillan, London, 1899–1930.
Golding, Harry (ed.), *The Wonder Book of Soldiers*, Ward Lock, London, various editions 1905–1940.
Grant, George, *The New Highland Military Discipline of 1757*, Museum Restoration Service, Ottawa, 1967.
Haswell-Miller, A.E. and Dawnay, N.P., *Military Drawings and Paintings in the Royal Collection*, 2 Vols, Phaidon, Oxford, 1966 and 1970.
Haythornthwaite, Philip, *Weapons and Equipment of the Napoleonic Wars*, Blandford Press, Poole, 1979.
Index to British Military Costume Prints, Army-Museums Ogilby Trust, 1972.
Kipling, Arthur and King, Hugh, *Head-Dress Badges of the British Army*, Frederick Muller, London, 1972.
Lawson, C.C.P., *A History of the Uniforms of The British Army*, 5 Vols, Peter Davies, Kaye and Ward, London, 1940–1967.
Luard, Lieutenant-Colonel John, *A History of the Dress of the British Soldier*, William Clowes, London, 1852.
Mollo, John, *Military Fashion*, Barrie and Jenkins, London, 1972.
Mollo, John, *Uniforms of the American Revolution*, Blandford Press, Poole, 1975.
Mollo, John, *Uniforms of the Seven Years War*, Blandford Press, Poole, 1977.
Nevill, Ralph, *British Military Prints*, Connoisseur, London, 1909.
Parkyn, Major H.G., *Shoulder-Belt Plates and Buttons*, Gale and Polden, 1956.
Robson, Brian, *Swords of the British Army*, 1788-1914, Arms and Armour Press, London, 1975.
Rogers, Colonel H.C.B., *Weapons of the British Soldier*, Seeley Service, London, 1960.
Rogers, Colonel H.C.B., *Battles and Generals of the Civil Wars, 1642–51*, Seeley Service, London, 1968.
Simkin, Richard, *Our Armies*, Simpkin, Marshall, n.d.
Stadden, Charles, *Coldstream Guards, Dress and Appointments, 1658–1972*, Almark, New Malden, 1973.
Strachan, Hew, *British Military Uniforms, 1768–96*, Arms and Armour Press, London, 1975.
Wace, Alan, *The Marlborough Tapestries at Blenheim Palace*, Phaidon, Oxford, 1968.
Walton, Colonel Clifford, *History of the British Standing Army, 1660–1700*, Harrison, 1894.
Wilkinson-Latham, John, *British Military Swords*, Hutchinson, London, 1966.
Wilkinson-Latham, R.J., *British Military Bayonets*, Hutchinson, London, 1967.
Wilkinson-Latham, R.J., *The Home Service Helmet, 1878–1914*, Collectors Series, n.d.
Wolseley, Viscount, *The Soldier's Pocket Book*, Macmillan, London, 1869 and 1886.
Young, Brigadier Peter, *Edgehill, 1642*, Roundwood Press, Warwick, 1967.
Young, Brigadier Peter, *Marston Moor, 1644*, Roundwood Press, Warwick, 1970.

REGIMENTAL HISTORIES

Works of this nature are too numerous to list here but special mention must be made of the following, which have useful and reliable appendices on dress:

Cavendish, Brig-Gen A.E.J., *The 93rd Sutherland Highlanders*, private 1928.

Cowper, Colonel L.I., *The King's Own; The Story of a Royal Regiment*, Oxford University Press, Oxford, 1939.

Leask, J.C. and McCance, H.M., *Regimental Records of the Royal Scots*, Dublin, 1915.

Leask, J.C., *Historical Records of the Queen's Own Cameron Highlanders*, London and Edinburgh, 1909.

Milne, S.M. and Astley-Terry, Major-General, *Annals of the King's Royal Rifle Corps — Appendix on Uniforms, Armament and Equipment*, London, 1913.

OFFICIAL PUBLICATIONS

Horse Guards Circulars, General and Army Orders.
King's and Queen's Regulations.
Royal Warrants.
Clothing Regulations.
Officers Dress Regulations.

PERIODICALS

Annual Reports, National Army Museum, London.
The Graphic.
The Illustrated London News.
The Illustrated Navy and Military Magazine.
Journal of the Royal United Services Institute.
Journal of the Society for Army Historical Research.
The King and His Army.
The Navy and Army Illustrated.
Soldier Magazine.
The United Service Magazine.
Plus various regimental journals.

MANUSCRIPTS

Lawson, C.C.P., *Notebooks*, National Army Museum, London.

Reynolds, P.W., *Military Costume in the 18th and 19th Centuries*, Victoria and Albert Museum, London.

ARTISTS

The artists and illustrators whose military work has proved most valuable in the compilation of text and coloured plates for this book are the following. They are given in roughly chronological order, from the seventeenth century to the present:
Jan Wyck, Marcellus Laroon, Louis Laguerre, John Wootton, Bernard Lens, the anonymous hand of the 1742 Clothing Book, Thomas and Paul Sandby, David Morier, J. S. Copley, P. de Loutherbourg, Edward Dayes, Robert Home, Edmund Scott, J. A. Atkinson, Charles Hamilton Smith, Denis and Robert Dighton, E. Hull, A. J. Dubois Drahonet, M. A. Hayes, L. Mansion and S. Eschauzier, Henry Martens, Thomas Baines, David Cunliffe, William Sharpe, G. H. Thomas, James Ferguson, Orlando Norie, Richard Simkin, Charles Fripp, Harry Payne, P. W. Reynolds, D. Macpherson, C. C. P. Lawson, A. E. Haswell Miller, Charles Stadden.

PICTURE CREDITS

Photographs and illustrations are supplied by or are reproduced by kind permission of the following:

The pictures on pages 12, 25 (bottom), 33, 36, 37, 38, 64, 72, 73, 74 and 88 are reproduced by gracious permission of HM the Queen.
The Armouries, H.M. Tower of London: 11; National Army Museum: 15, 18, 19, 20, 21, 26, 27, 28, 32, 39, 44, 47, 48, 49, 50, 51, 57, 60, 61 (bottom), 62, 63, 68, 75, 76, 82, 83, 85, 92, 93, 99, 101, 108, 111, 113, 116, 119 (top), 120, 124, 126, 128, 140; Imperial War Museum: 129, 133; British Museum: 16; National Galleries of Scotland: 45 (right); Victoria & Albert Museum: 53, 54, 55, 56 (top); Sotheby, Parke, Bernet & Co: 25 (top); M. Knoedler & Co: 29; A. C. Cooper: 41 (left); *SOLDIER* magazine: 135 (top); Prince Consort's Library, Aldershot: 40; Army Museums Ogilby Trust: 45 (left); Defence NBC Centre: 142; Worcestershire and Sherwood Foresters Regiment: 61 (top); Queen's Own Highlanders: 84; Rifle Brigade Museum Trust: 89; 7th Gurkha Rifles: 136; Private Scottish Collection: 41 (right); D. S. V. Fosten Collection: 119 (bottom); R. G. Harris Collection: 95, 105, 106, 110; R. J. Marrion Collection: 96, 98, 107, 112; Denis Quarmby Collection: 94; Charles Stadden Collection: 135 (bottom); Mathew Taylor Collection: 97; Author's Collection: 56 (bottom), 81, 107 (top), 117, 127, 134.

Index

Figure numbers in italic refer to colour illustrations: page numbers in italic refer to black & white illustrations.

REGIMENTAL INDEX

All Guards and all Line regiments numbered 1st — 96th are listed between pp 144–147.
All Line regiments from 1st/Royal Scots to 109th/2nd Leinster and the Rifle Brigade, as well as all existing regiments with administrative divisions, are listed between pp 150–151.
All regiments listed below are in order of precedence conferred by pre-1881 numbers: all amalgamated/large regiments formed since 1958 take precedence of their senior former regiment.